Advances in Military Geosciences

Series Editors

Military activities are almost always strongly integrated within a wide spectrum of geoscience. The decisive outcomes of land battles throughout history have been dictated in large part by the terrain and environmental setting. Modern military operations rely on a wide range of land-, air-, sea-, and space-borne intelligence and knowledge of dynamic terrain processes and conditions. In addition, the study of geo-based environmental science is critical to both the sustainable management of military reservations and installations, as well as the evaluation of how terrain and environmental conditions may impact military equipment and operations.

Advances in Military Geosciences contains single and multi-authored books as well as edited volumes. Series Editors are currently accepting proposals, forms for which can be obtained from the publisher, Ron Doering (ron.doering@springer.com).

More information about this series at http://www.springer.com/series/15030

Francis Galgano
Editor

The Environment-Conflict Nexus

Climate Change and the Emergent National Security Landscape

Springer

Editor
Francis Galgano
Department of Geography and the Environment
Villanova University
Villanova, PA, USA

ISSN 2522-8315 ISSN 2522-8323 (electronic)
Advances in Military Geosciences
ISBN 978-3-030-08146-1 ISBN 978-3-319-90975-2 (eBook)
https://doi.org/10.1007/978-3-319-90975-2

Softcover re-print of the Hardcover 1st edition 2019

Printed on acid-free paper

This Springer imprint is published by the registered company Springer International Publishing AG part of Springer Nature.
The registered company address is: Gewerbestrasse 11, 6330 Cham, Switzerland

Preface

The evolution of the global strategic situation following the Cold War suggests the need to expand the definition of national security to include environmental threats to stability. During the past two decades, there has been a dramatic shift in how we perceive the contemporary national security landscape. Leaders of governmental organizations and nongovernmental agencies have progressively come to accept that the harmful effects of climate change and other environmental factors are exposing vulnerable societies to instability and potentially, violent conflict. This altered perception of the linkages among global environmental problems and related economic and demographic challenges has now emerged as one basis for interpreting conflict and security. Certainly, the Intergovernmental Panel on Climate Change (IPCC) has dedicated a great deal of effort to assessing the vulnerability of human populations resulting from exposure to the adverse effects of climate change. The Fourth Assessment Report of the IPCC, which examines the issue of exposure, adaptation, and vulnerability, suggests that counties and societies, especially in the developing world, will have difficulty adapting to the strain of climate change in the not too distant future. Adaptation and resilience will be hindered by a lack of capacity, and the people hardest hit will be those living in poverty and within failed states.

This book is about the link between the environment and conflict. Environmental security refers to a range of security issues triggered or intensified by environmental factors such as climate change, resources, demographic factors, natural disasters, environmental change, and nonsustainable practices. Environmental stress has the potential to destabilize states, especially in the developing world because they are characteristically more dependent on the environment for economic productivity and they lack the resiliency to overcome these challenges. This perspective has considerably refocused the lens by which we view the environment as a variable in the national security calculus. As population and economic demands escalate, and the adverse effects of climate change become more apparent, collectively these problems may disrupt vulnerable populations to the extent that they erode governmental legitimacy, thus making them more vulnerable to instability and conflict.

Some dispute the relationship between the environment and conflict and suggest that violent conflict results only from political and military factors. Clearly, it is difficult to identify conflicts in which environmental conditions are the causative factors. However, while the details of a potential conflict triggered by environmental factors cannot be predicted, the historical record provides useful guidelines because the evidence is clear that this linkage exists. The environmental security perspective given in this book does not assert that the nature of conflict is new; rather, it suggests that because environmental stress is growing worse, we can expect an increase in the frequency of conflicts with an environmental component. Additionally, the effects of climate change are not restricted by state boundaries. Indeed, research presented in this book demonstrates that developed and developing states are vulnerable to instability. However, data clearly suggest that the problem is spatially concentrated and greatly magnified in the developing world. These states are more vulnerable because they suffer from several persistent environmental and human variables such as environmental degradation, reduced agricultural production, economic decline, poor governance, population growth and displacement, and civil disruption.

Clearly, identifying states at risk to instability and violence from environmental causes involves an extensive and complex array of security issues, particularly if we define it very broadly. This book, however, narrows considerably the scope of environmental security by focusing exclusively on how the environment affects conflict: i.e., the environment–conflict nexus. In so doing, the book offers a series of case studies that examine this nexus from a variety of perspectives (e.g., water, climate change, urban areas) and from different scales (i.e., local to global).

This book begins with three chapters that set the stage for the case studies that follow. In the first chapter, Francis Galgano establishes the importance of environmental factors on the emergent national security landscape, and in the second, he presents a quantitative index to identify states vulnerable to violent conflict resulting from exposure to the adverse effects of the environment. In the third chapter, Adam Kalkstein defines the scope of climate change and its influence on the environmental security model. The three opening chapters are followed by case studies that examine the environment–conflict nexus from a variety of perspectives and scales. The first such case study is presented in the fourth chapter in which Francis Galgano examines the global environmental disaster that may be precipitated by an abrupt climate change. This is a global-scale projection, based on plausible evidence supported by actual abrupt events that have occurred in the Holocene climate record. In the fifth chapter, Francis Galgano examines the global problem of renewable water resources and transboundary watersheds with a regional focus on the problem of water in the Middle East. This region has the world's fastest growing population and its renewable freshwater resources are strained beyond sustainable levels. In the next chapter, Dr. Amy Richmond examines environmental security from the perspective of problematical and nearly uncontrolled urban growth in sub-Saharan Africa. In the seventh chapter, Wiley Thompson examines the problem of Chinese expansion into the South China Sea through their practice of developing coral reefs into military bases, and expanding territorial claims. In the eighth chapter, Andrew Lohman presents an historical vignette from the First World War. In this case study, he describes the military campaign in East Africa, which was part of the larger imperialist roots of the war. In the ninth chapter, Francis Galgano presents a case study of the Ogaden War of 1977, which illustrates a conflict that was triggered by exposure to a climate shock, in this case a decade-long drought. In the final chapter, Amy Richmond and Francis Galgano assess the root causes of the conflict and resultant genocide in Rwanda from an environmental security perspective. In the final chapter, Mark Read examines linkages between climate change, drought, migration, and the civil war in Syria.

The environment–conflict nexus has been propelled in large measure by globalization, which has eliminated much of the friction of distance and created expectations in the developing world of economic growth and affluence. It has also accelerated economic demands, leading to unsustainable economic activity and environmental damage, which combined with population pressure and climate change has stressed many ecosystems beyond their capacity. The dynamics of globalization have contributed to the number of failing and failed states incapable of keeping pace with the demands of environmental change, thus creating ungoverned spaces ripe for instability and conflict. As global population grows, economic demands may exceed the natural resource and economic base of many states, erode governmental legitimacy, and promote intrastate conflict over increasingly scarce resources. This topic is of considerable importance because the geopolitical implications of environmental security for US foreign policy are significant. With conflicts and recent environmental disasters in Rwanda, East Timor, Haiti, Darfur, and Syria as the precedent, the use of Western and United Nations (UN) military forces to address humanitarian dimensions of regional conflict has been now well established. However, UN and Western leadership has approached these commitments with acute reluctance. Nonetheless, conflicts with an environmental component coupled with divisive ethnic dimensions have increased pressure on the West and UN to commit resources to stability efforts.

The scenarios presented in this book clearly indicate that the future is not bright given our profound alteration of the natural environment and the weakening of government control in many states. Fortunately, however, this bleak prognosis is only a forecast based on contemporary trends. Like all predictions, it is rooted in contemporary trends and recent past history, and they do not, however, necessarily reflect a viable image of the future. Human society is not predestined to enter into an agonizing decline into environmental chaos, and clearly, there are scientific, technical, and economical solutions that can reduce the level of environmental stress and diminish potential conflict—and there are important social institutions that promote stability over chaos. However, these opportunities are opposed by considerable social, political, and institutional barriers. In order to prevail and lower the threshold of environment threats, the global community must deal with the roots of environmental instability.

Villanova, PA, USA Francis A. Galgano

Contents

Contents

About the Editor

Francis A. Galgano received a B.S. from the Virginia Military Institute in 1980. He later earned his master's and doctoral degree from the University of Maryland at College Park in 1989 and 1998, respectively. He retired from the Army as a Lieutenant Colonel after 27 years in 2007—at that time, he was serving on the faculty of the United States Military Academy. He is presently the Chair of the Department of Geography and the Environment at Villanova University. Dr. Galgano has coedited three military geography books and has authored a number of publications on various geographical subjects.

Contributors

Francis A. Galgano Department of Geography and the Environment, Villanova University, Villanova, PA, USA

Adam J. Kalkstein Department of Geography and Environmental Engineering, Center for Languages, Cultures, and Regional Studies, United States Military Academy, West Point, NY, USA

Andrew J. Lohman Department of Geography and Environmental Engineering, United States Military Academy, West Point, NY, USA

Mark R. Read Department of Geography and Environmental Engineering, United States Military Academy, West Point, NY, USA

Amy K. Richmond Department of Geography and Environmental Engineering, United States Military Academy, West Point, NY, USA

Wiley C. Thompson Oregon State University Extension Service, Eugene, OR, USA

The Environment–Conflict Nexus

Francis A. Galgano

Abstract The evolution of the global strategic situation following the Cold War suggests the need to expand the definition of national security to include environmental threats to stability. Environmental security refers to a range of security issues triggered or intensified by environmental factors such as climate change, resources, demographic factors, natural disasters, environmental change, and non-sustainable practices. Environmental stress has the potential to destabilize states, but especially in the developing world because they are characteristically more dependent on the environment for economic productivity and they lack the resiliency to overcome these challenges. This chapter does not suggest that environmental stress—alone—causes warfare. To be more precise, it can potentially trigger violent conflict in unique situations of extreme civil instability and within failing states. The problem that we face today is that the number of failing states is growing, and they are more vulnerable to instability caused by environmental stress because they suffer from four causally related effects: (1) reduced agricultural production; (2) economic decline; (3) population displacement; and (4) civil disruption. These effects determine the vulnerability and adaptability of a society.

Keywords Climate change · Cold War · Conflict · Demographics · Desertification · Drought · Environmental Change · Environmental Determinism · Environmental Security · Environment–Conflict Nexus · Failed States · Globalization · Governance · Internecine Conflict · Malthus · National security · NATO · Population · Poverty · Regional Warfare · Soil Erosion · Warsaw Pact · Water

F. A. Galgano (✉)
Department of Geography and the Environment, Villanova University, Villanova, PA, USA
e-mail: francis.galgano@villanova.edu

F. Galgano (ed.), *The Environment-Conflict Nexus*, Advances in Military Geosciences, https://doi.org/10.1007/978-3-319-90975-2_1

1 Introduction

Since the end of the Cold War, the perception of the national security landscape has evolved and linkages between the environment, political instability, and violent conflict—that is, environmental security—have become an increasingly accepted paradigm in security affairs. Environmental security refers to a broad range of security issues triggered or exacerbated by demographic and environmental factors such as competition for resources, population growth and displacement, disease, natural disasters, climate and environmental change, resource shortages, and non-sustainable practices (Harnish 2009). During the past two decades, there has been a shift in governmental circles and well as the academic community's perception of global environmental problems and their link to destabilizing societies (Solow 2011; Femina and Werrell 2012). Indeed, the Intergovernmental Panel on Climate Change has dedicated a great deal of effort to assessing the vulnerability of human populations resulting from exposure to the adverse effects of climate change (IPCC 2007, 2012, 2014). Thus, the environment has emerged as one basis for interpreting conflict and security. This is made more complicated because the environmental security paradigm encompasses an extensive and complex array of security issues, particularly if we define security very broadly to include societal, environmental, social, and economic wellbeing. Consequently, this book and the chapters herein, focus on the environment–conflict nexus, which is defined as political instability and violent conflict enabled by the exposure of a vulnerable population to the adverse effects of the environment.

Hence, this book examines linkages between environmental stress, political instability, and conflict; and the analysis provided in its chapters suggests that developing states are more vulnerable because they suffer from several persistent and causally–related factors, such as environmental degradation; reduced agricultural production; economic decline; weakening governance; population growth and displacement; and pervasive civil disruption. These problems are magnified because of the persistent and problematical adverse effects of global climate change (IPCC 2014). Furthermore, the dynamics of globalization has eliminated the friction of distance and created expectations in the developing world of economic growth, thus intensifying the gap between developed and developing states (Butts 2011). Although, the IPCC (2012) indicates that all countries are vulnerable to climate change, developing states are consistently more vulnerable. This sharpens the lens by which we view the environment as a variable in the national security calculus. As populations grow and economic demands increase, and the adverse effects of climate change manifest themselves within states already struggling with governance issues, the combined effects of these problems may exceed the natural resource and economic base of the state and erode governmental legitimacy, thus making them more vulnerable to conflict (Smith and Vivekananda 2009).

The prevalence of the environment–climate nexus also suggests that continued peaceful resolution of environmentally triggered conflict is inconsistent with the realities of the emerging national security landscape. Given these circumstances, it is plausible that we will witness a surge in three modes of conflict related to the environment–conflict nexus: internecine conflict driven by environmental stress and demographic trends; civil war prompted by governmental collapse and/or economic failure; and limited-scale interstate conflicts. This assessment is related to three persistent realities. First, climate change is magnifying extant demographic and environmental factors beyond the adaptive capacity of many states. Second, the proliferation of failing states has singularly reduced resilience and the potential for diplomatic resolution in many regions. Finally, competition for essential resources has been exacerbated by population growth and globalization in many regions (Yohe et al. 2006). Thus, I argue that environmental factors will likely provide a tipping point that advances violent conflict in regions that may already be on the brink of instability.

The environment–conflict nexus has engendered particular concern in U.S. government circles. In the 2014 Quadrennial Defense Review, the Department of Defense indicated that,

> … pressures caused by climate change will influence resource competition while placing additional burdens on economies, societies, and governance institutions around the world. These effects are threat multipliers that will aggravate stressors abroad such as poverty, environmental degradation, political instability, and social tensions – conditions that can enable terrorist activity and other forms of violence (DoD 2014, p. 30).

Thus, with environmentally related conflicts and humanitarian disasters in Somalia, Rwanda, East Timor, Haiti, Banda Ache, Syria, and Darfur as the precedent, the use of Western and United Nations (U.N.) resources and military force to address humanitarian dimensions of regional conflict has been now well established, and it appears that environmental change and resource scarcity may already be contributing to instability and violence (Solow 2011). A word of caution however, the environmental security perspective does not assert that the nature of conflict is new and I do not hypothesize that the causal links between environmental variables and conflict are deterministic. Rather, I propose that potential conflict related to environmental factors cannot be predicted accurately—I do, however, suggest that we can determine which states are most vulnerable given a set of variables.

The environment–conflict nexus encompasses a broad set of factors that endanger human security; and many anthropogenic processes combine with natural processes environmental conditions to enable instability resulting from ignorance, accident, mismanagement, or design (Hsiang et al. 2011). Yet, the problem is that delineating factors that contribute to environmental instability is an inexact method involving environmental risk analysis based on complicated linkages between human and natural processes. Therefore, it is helpful to establish a framework—or model—to delineate the various factors that are operating in a region and from which cogent analyses can be made.

2 The Environment–Conflict Nexus

> Few threats to peace and survival of the human community are greater than those posed by the prospects of cumulative and irreversible degradation of the biosphere on which human life depends. … Our survival depends not only on military balance, but on global cooperation to ensure a sustainable environment. Brundtland Commission Report, U.N. (1987)

This book is focused on the environment–conflict nexus because contemporary events suggest that there is a link between the environment and conflict. States are susceptible to this nexus because exposure to the adverse effects of environmental change can destabilize governments and societies, thus making them increasingly vulnerable (Wagner 2005). However, the link between the environment and conflict is a matter of some polemic and continues to inspire debate in academic and professional circles. Nevertheless, contemporary research suggests that climate and environmental factors are already contributing to political instability and violence (Bennett 1991; Shah and Landay 2010; Burke et al. 2009; Solow 2011). At the crux of the matter are three critical and interrelated factors. First, the adverse effects of climate and environmental change are having a more pervasive and debilitating effect on people and governments, thus eroding their ability to adapt (IPCC 2012). The second is governance, the number of failing states is growing and adaptive capacity and stability is tied strongly to governance (Smith and Vivekananda 2009). Failing states are problematical because they have large areas that are outside of effective government control and are thus affected severely by humanitarian disasters, environmental stress, and internecine conflict (Galgano 2007). The third factor is economic. Poverty at national and household levels intensifies vulnerability to environmental stress and degrades resilience (Hendrix and Salehyan 2012). This dilemma is expected to grow worse during the coming decades given that global population will exceed 9.0 billion and to keep pace, economic output will have to quintuple (Homer–Dixon 1991).

Environmental stress is having a fundamental effect on stability because the economic welfare of more than 3.5 billion people—about half of the world's population—is tied to the land. Therefore, factors such as agricultural productivity, water, fuel, and deforestation are crucial environmental indicators; especially given the dual problems of population growth and climate change (IPCC 2007). Drought, desertification, deforestation, soil erosion, and exhaustion are major problems in many regions, but especially in the developing world; where exposure to the adverse effects of environmental change is of great consequence because almost 75% of the world's most impoverished inhabitants are subsistence farmers facing declining productivity (Mutunga et al. 2012). These dynamics have important security consequences and represent the potential to undermine states that lack the resource base, institutional strength, and resiliency to meet these challenges.

Nevertheless, it is atypical for linkages between the environment and conflict to be directly and absolutely causative. Although, in many examples environmental phenomena contribute to conflicts, they are seldom the sole causes. Ongoing research, and indeed the chapters in this book, indicate that there are too many other variables to consider, such as social stratification, weak economies, and repressive governments by way of example. Each of these variables could destabilize a society, but in developing countries where absolute poverty, ineffective governance, the absence of reliable shelter, and pervasive health issues afflicts about 1.3 billion people, the adverse effects of environmental degradation and climate change can clearly serve as the trigger to violent conflict. History has demonstrated that impoverished people become desperate and all too ready to resort to force to overthrow governments or secure the resources they see as being necessary to their survival. Furthermore, especially in failing states, the environment–conflict nexus can stimulate the use of force by the government to repress disaffection among those who suffer the consequences of environmental decline.

Hence, environmental deficiencies and the effects of climate change create circumstances within which conflict is more likely: they can affect the character of conflict; they can determine the source of conflict; and they can act as multipliers that aggravate core causes of conflict. However, to reiterate a fundamental point—there are usually a number of factors that undermine security. They include faulty economic policies, inflexible political structures, oligarchical regimes, oppressive governments and other adverse factors that have nothing directly to do with environment. Nevertheless, these deficiencies typically exacerbate environmental conditions, and are aggravated, in turn, by environmental problems.

This is important because, the U.N. and Western leadership have approached these challenges with acute reluctance; nonetheless, conflicts with an environmental component have increased pressure on the West and U.N. to commit resources to stability efforts (Dulian 2004; Drapeau and Mignone 2007). Thus the seminal question, especially given the anticipated effects of climate change, is can the adverse effects of the environment destabilize a state and enable violent conflict; and effectively change the national security calculus?

2.1 *The Evolving Security Landscape*

The environment–conflict nexus has generated increased interest in professional and academic circles since the mid–1980s. The broader contemporary national security debate suggests that the potential for violent conflict triggered by environmental stress looms over society, which is much different from the traditional Cold War concept of security (Femina and Werrell 2012). Thus, a shift has taken place: during the Cold War, conflict and alliances formed almost exclusively along political lines; but now we have begun to pay greater attention to problems evolving from intensified competition over essential resources and environmental degradation (Floyd 2014). Environmental security first emerged as a potential variable on the security landscape during the mid–1980s; but it did not become firmly established until 1987, notably through the publication of the so–called *Brundtland Report,* which stated that. "*... environmental stress is both a cause and an effect of political tension and military conflict,*" (U.N. 1987, p. 290). This was followed by a hiatus in professional and academic studies; however, it attracted renewed interest during the mid–1990s in governmental security documents, but especially following the fall of the Soviet Union because environmental security represented a fundamental change in the interpretation of national security affairs (Galgano 2013).

The environment first became an element in the U.S. National Security Strategy in 1988 when President Reagan's National Security Strategy identified threats to the U.S. from the Soviet Union's nuclear arsenal, but also from the environmental perspective "*the dangerous depletion or contamination of natural endowments of some nation's soil, forest, water, and air*"...which, "*create potential threats to the peace and prosperity that are in our national interests as well as the interests of the affected nations*" (White House 1988, p. 6). Later National Security Strategies followed suit by suggesting that the environment was a potential trigger for violent conflict (White House 1991, 1997). For example, the National Security Council (NSC) pointed out that, "*... stress from environmental challenges is already contributing to political conflict,*" (NSC 1991, p. 2). The 1991 National Security Strategy further indicated that it was a primary U.S. objective to "*... achieve cooperative international solutions to key environmental challenges,*" (NSC 1991, 21).

By 2005, the U.S. Department of Defense (DoD) identified environmentally related instability as a fundamental strategic concern, and that environmentally–triggered conflict typically manifests itself in failing states, thus making its international management and intervention difficult (DoD 2005). In his 2010 National Security Strategy, former President Obama reinforced the link between the environment and conflict when he listed environmental factors and resource scarcity as important features of the security landscape. He indicated that conflicts driven by ideology might give way to conflict triggered by demographic and environmental factors, "*Wars may no longer simply be about armies and weapons ... rather, it increasingly correlates to environmental factors and dynamics that have been rarely considered by national leaders*" (Obama 2010, p. 4). Therefore, national security affairs may no longer only be about traditional politico–military dynamics; rather, climate, resources, and demographics may now be viewed as being equally important as traditional elements of national power. In the 2014 Quadrennial Defense Review, the DoD clearly indicated that it viewed the environment as a contributor to regional instability and violence, "*Competition for resources, including energy and water, will worsen tensions in the coming years and could escalate regional confrontations into broader conflicts – particularly in fragile states*" (DoD 2014, p 14).

The environment–conflict nexus has been increasingly recognized in academic circles as well. Homer–Dixon (1991) proposed a conflict causality model that linked the environment to conflict and suggested that failing states are more vulnerable to environmental stress and suffer from four fundamental causally related effects: 1) reduced agricultural production; 2) economic decline; 3) population displacement; and 4) civil disruption (Homer–Dixon 1991). Kaplan (2000) suggested that environmental factors represent the core foreign policy challenge in this century, and indeed, the ongoing discourse regarding the potential security and political implications of climate change has promoted environmental security to the forefront of the global security agenda (Maas et al. 2014). Klare (2001) suggested a national security geography to explain the evolving spatial dynamics of conflict following the Cold War—this one driven by competition over vital resources. Smith and Vivekananda (2007) examined the nexus of environmental stress and failing states and their analysis suggests that there are 46 developing states (2.7 billion people) within which the effects of climate change coupled with weak governance will create a high risk of violent conflict by the end of this century. Burke et al. (2009) conducted a comprehensive examination of global climate change and its potential linkages to armed conflict in sub–Saharan Africa, and suggest that there will be a 54% increase in the incidence of armed conflict by 2030. Hsiang et al. (2011) developed a quantitative model using ENSO data from 1950–2004 and demonstrated that the probability of conflict doubles in the tropics during El Niño years. Hendrix and Salehyan (2012) examined deviations from normal rainfall patterns in Africa and their results indicate that extreme variations in precipitation are associated positively with political and civil conflict.

Thus, it appears that environmental change and resource scarcity is already contributing to instability and violence, but especially in the developing world (Solow 2011). The environment–conflict nexus is a phenomenon that is correlated to low levels of economic development and high levels of agricultural dependence (Hendrix and Salehyan 2012). Links between environmental effects and conflict appears to be strongest in less developed states because the inability to adapt fosters grievances among disenfranchised groups and encourages heightened competition for natural resources. Hence, conflict is enabled in these scenarios because factors related to social stratification, which includes poverty and poor governance, make them more vulnerable (Solow 2011). Although there is growing evidence of the potentially disruptive effects of environmental change, the critical problem is defining the tipping point between societies that can adapt and those with highly stressed environments that cannot (Burke et al. 2009).

The problem facing the West is that, in the developing world, the capacity to adapt is declining as governments continue to fail and are simultaneously stressed by climate change (Galgano 2007; IPCC 2012). Consequently, the concept of environmental security has emerged as one basis for understanding conflict and security, and two broad doctrines have materialized. Traditionalist desire to confine the subject of conflict to politico–military dynamics, while others desire to broaden the discipline to embrace the environment and its potential effect on conflict (Foster 2001). The latter group certainly does not maintain that the character of modern conflict is somehow fundamentally different or unique. Rather, they suggest that because environmental stress is worsening and the number of failing states is growing; we can expect a surge in the frequency of conflicts such as, insurgency, ethnic clashes, civil war, and revolt with an environmental component (Bennett 1991; Klare 2001; Knickerbocker 2007; Sappenfield 2007; Butts 2011; Krakowka 2011).

2.2 A Malthusian Link?

Detractors of the environment–conflict perspective argue that conflicts result exclusively from politico–military factors, and are rarely induced by the adverse effects of environmental change, and hint at environmental determinism and the weakness of the Malthusian argument (Foster 2001). However, the environment may play a variety of roles in triggering conflict and environmental security doctrine is only one plausible explanation, not a deterministic model. Environmental stress and resource scarcity result from the combined influence of anthropogenic effects on the environment in conjunction with natural processes, and the sensitivity of the ecosystem. Consequently, environmental factors contribute to conflict only under particular conditions—there is no deterministic link between these variables (Percival and Homer-Dixon 1995). Clearly, not all conflicts are identical and the potential influence of environmental stress on instability and warfare will vary in magnitude from situation to situation (Krakowka 2011). Though the environment–security paradigm suggests that environmental change can enable violent conflict, studies have not been able to establish a clear causal link between the two. A great deal of evidence appears to be anecdotal and links are speculative, which is a fair criticism of the paradigm; and it also underscores the principal weakness of the environmental security model, which is a lack of predictive capacity (Solow 2011). That is, we have no overarching appreciation of which environmental scenarios will lead to conflict. While we do not have a complete picture of the causal links within the environment–conflict nexus, the old maxim that, "*... the absence of evidence is not an evidence of absence*," is a suitable reminder that we can at least accept the plausibility of the paradigm; and that violent conflicts could ensue for a variety of causes that become more probable when environmental conditions deteriorate (Femina and Werrell 2014, p. 2).

Dynamics between population, governance, resources, economic base, environmental stress, and conflict are very complex and not a simple deterministic recipe. The outcome of a potential environmental security scenario is influenced strongly by government policy, social structure, strength of governance, technology, and infrastructure. The relationships between food, population growth, and environmental stress are evident in many developing states; nonetheless, this Malthusian paradigm generates much disagreement among researchers. However, regardless of the perspective, there appears to be one irrefutable outcome: that is, environmental stress superimposed over underlying societal–political divisions will be resolved—one way or another. Historical events indicate, unfortunately, that their resolution sometimes can be violent. The challenge is the ability to develop an analytical framework to detect the tipping point between a society's resiliency and adaptability, and chaos and violence.

3 Contemporary Challenges

The emergence of the environment–climate nexus is in large measure a function of key challenges that are part of the contemporary national security landscape, coupled with the pervasive effects of global environmental change. The Cold War strategic partition of the world dominated the global security landscape for more than four decades and fostered a rather simplified national security geography. This bipolar world was dominated by two super-powers and their respective alliance blocks (i.e., NATO and the Warsaw Pact). However, a major tectonic shift in the geography of national security occurred following the collapse of the Soviet Union and dismemberment of the Warsaw Pact in 1990, characterized by redefined regional balances of power, new conflict dynamics, and environmental challenges, all of which presented a new set of geographic realities. The "old" national security map was dominated by major power blocks, but at its core, the state remained the predominant integer in the global strategic calculus. This strategic geography focused on the nuclear balance-of-power and conflict was limited to regional hot zones triggered by surrogate regimes. Nonetheless, the core of the former national security landscape was the power of the state and pervasive superpower influence, which ensured that adequate control was exerted over proxy regimes to preclude regional hot spots from escalating into major violent confrontations. The fundamental national security dilemma of the modern age is that we now live in a multi-centric word—that is national security is influenced nearly equally by traditional states and transnational actors; and the problems of governance, environmental change, and population growth have manifested themselves by triggering regional instability (DoD 2014).

Hence, the emergent national security landscape is far more complex and much less stable. Contemporary national security is dominated by many more problematical factors, such as globalization and the spread of technology; the proliferation of nuclear and non-nuclear threats; environmentally triggered instability; instability generated by weakly governed states; and the emergence of violent non-state actors (Femina and Werrell 2014). This new national security landscape is very different and far more unstable because, notwithstanding the danger inherent to the "old" national security map, it was nonetheless state-centric. Given that the state was the dominant actor, conflict resolution and global balance could be achieved by quasi-peaceful means using established diplomatic and international protocols. In such cases, well-established diplomatic doctrine was reasonable and viable. However, the emergent geography of national security is influenced strongly by violent non-state actors and the overarching problem of effective governance. In other words, many states cannot, or will not exert effective control over their territory, and thus, it appears that the old rules of state sovereignty may no longer apply.

President Barak Obama defined the United State's national security policy in his *National Security Strategy* (White House 2010). In articulating the U.S. national security strategy posture, he acknowledged the emerging threats that form the underpinnings of the global security environment: the spread of deadly technologies, terrorism, economic upheaval, and instability enabled by environmental

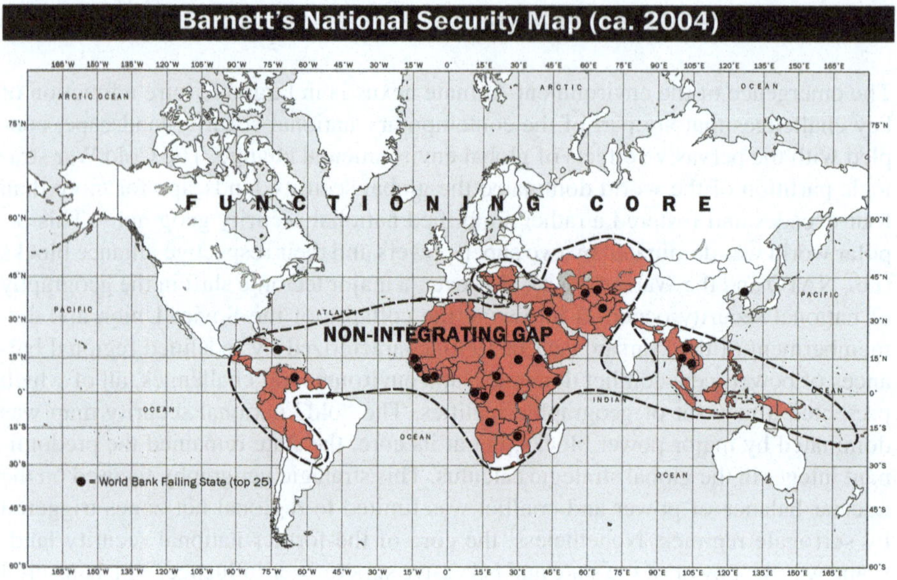

Fig. 1 Evolution of the post-Cold War security landscape. Barnett's (2004) map illustrates the separation of the non-integrating gap from the rest of the world. Black dots on the map indicate the 25 least effectively governed states in the world. Cartography by the author.

change. These threats, in addition to traditional state-centric threats, are enabled by contemporary dynamics such as globalization, emergence of violent non-state actors, proliferation of failed states, and effective sovereignty issues.

Thomas Barnett (2004) developed a new national security geography (Fig. 1) that takes into consideration the concept of a closed system, however, he has changed the prevailing view and incorporated emerging post-Cold War dynamics that dominate the global security calculus: i.e., globalization, failing states, and violent non-state actors. In Barnett's view, these factors are linked and are destabilizing large segments of the world. Furthermore, these dynamics are forcing a recasting of U.S. national security doctrine.

In Barnett's estimation globalization of the economy during the 1990s, and the boom that followed did not lead to a new era of integration and world peace as many expected. In fact, the uneven nature of economic prosperity nurtured by globalization and the hostility engendered by its related cultural assimilation—especially in the Islamic world—has made it possible for violent non-state actors to operate as never before. This suggests a seismic shift in national security policy because Barnett's map (Fig. 1) presents a world that is bifurcated into one that it actively integrating itself into a so-called Functioning Core, and one that is trapped in a Non-Integrating Gap. The Non-Integrating Gap is largely disconnected from the rest of the world, is characterized by failing states that are highly vulnerable to exposure to climate change. Dealing with and mitigating this disconnectedness, coupled with the adverse effects of climate change, has become the defining global security issue of our time (IPCC 2014). Barnett presented his theory in 2004, and it was essentially validated in President Obama's 2010 *National Security Strategy* within which the issues of governance, violent non-state actors, environmental change, and demographics factors called for new rules by which we define the national security landscape.

The overriding issue that is driving the environment–climate nexus is climate change and its adverse effects. The projected effects of climate change and its potential to destabilize states dwarf previous environmental problems and portend political, economic, and social disruptions on a global scale (IPCC 2014). Significant changes in global temperature and precipitation patterns could have catastrophic consequences for the habitability of entire regions, thus generating severe instability and potential for violent conflict. While broad scientific consensus exists on the underlying theory and effects of climate change, the nature and effects of global warming as predicted by computer models remains a matter of some polemic.

In many regions, exposure to climate change is affecting human and natural systems and the consequences are becoming more clear (IPCC 2014). In the context of environmental security, this variable is essential because exposure reduces resiliency, increases vulnerability, lowers the threshold for violent conflict. Exposure to the significant effects of climate change (i.e., water scarcity, lowered food production, health effects, population displacement, etc.) have a high correlation to violent conflict (Smith and Vivekananda 2007; Burke et al. 2009; Hsiang et al. 2011; Solow 2011; Femina and Werrell 2012).

3.1 Governance and Failed States

Identifying fragile and failing states is essential because they typically manifest persistent poverty, a lack of infrastructure, and poor governmental control, all of which degrade the ability of a society to adapt (Smith and Vivekananda 2007). Although the effects of climate change are not confined by state boundaries, the effects of environmental security are not randomly and fairly distributed because adaptive capacity is a function of the strength of the state and the financial resources of the population. Thus, the effects of environmental security are defined by state boundaries and governance is a key indicator. The IPCC (2014) suggests that the people most affected by the problematic effects of climate change are those living in poverty in which fragile governance and weak economies erode their livelihood and security. Environmental stress alone does not, inevitably, trigger warfare. Evidence suggests that it enables or intensifies violent conflict when it combines with weak governance and social division, to affect a spiral of violence, typically along ethnic and political divisions. Contemporary trends indicate that environmentally-driven violence has been concentrated in the developing world because it exhibits extreme social fragmentation and stratification. Developing states are more susceptible to environmentally triggered conflict because they are, characteristically, more dependent on the environment for their economic productivity; and lack the resiliency to overcome these challenges because they have weak economies and small capital reserves, shortages of scientists and engineers, and poor distribution infrastructure (Galgano 2007).

Governance is a critical problem in the developing world and since 1990, the number of failing states has grown. The World Bank (Kaufmann et al. 2010) examined governance by indexing six key metrics as a means to quantify state stability and rule of law. Their findings indicate that of 187 states examined, 92 exhibited problematic levels of instability: i.e., 49% of all states (Kaufmann et al. 2010). This is a troubling trend. First, government stability and effectiveness in developing regions such as Sub-Saharan Africa, South America, and Asia is growing weaker. Second, the rift between the developed and developing world is expanding. Finally, recent World Bank data indicate that, while governance scores have marginally improved, the number of problem states has increased. Of 212 state entities examined in 2016, 122 exhibit instability—i.e., 57% of all states (WGI 2016).

Failing states are troubling because they have large areas that are outside of effective government control and thus, can be affected severely by humanitarian disasters, environmental stress, and ethnic conflict (Galgano 2007). This is because they lack effective institutions and resources to safeguard the population. Thus, weak or failing states are more vulnerable to environmental stress and suffer from four fundamental causally-related social effects: (1) reduced agricultural production; (2) economic decline; (3) population displacement; and (4) civil disruption (Homer–Dixon 1991). These effects fundamentally determine the vulnerability and adaptability of the society. This raises the complexity of the problem for governments as well as non-governmental organizations and intergovernmental bodies as they attempt to develop relief strategies (Galgano 2007).

3.2 Water

Water scarcity is a leading indicator of environmental stress and instability because of the depletion of fresh water resources from population growth, environmental degradation, climate change, and conflict. Water scarcity is a variable expressed as renewable water resources (i.e., meters^3) *per capita*. In practical terms the average person requires, to survive, a minimum of 1 m^3 of water per year for consumption, 100 m^3 for domestic use, and 1000 m^3 to grow the food that is consumed. Thus, states with less than 1000 m^3 of water per capita are considered water–scarce, and those with less than 1700 m^3 per capita are considered water–stressed (U.N. 2009).

Water has long served as a key illustration of the environment–conflict nexus. For example, 25% of all water–related disputes during the past 50 years have resulted in some form of hostilities, and 37 of them have resulted in military action (Gleick 2012). Indeed, one of the motivating factors for the preemptive Israeli offensive during the 1967 Six Day War was the threat by Syria and Jordan to sequester a good part of the River Jordan's discharge—which Israel viewed as a sufficient *casus belli*. Similarly, Ethiopia has been asserting its plan to divert much of the Blue

Nile to irrigate extensive sectors of its highland region, which has caused Egypt to threaten Ethiopia with a declaration of war.

Since 1950 the global freshwater supply *per capita* has fallen by 60% as the global population has swelled by over 150%. Consumption can be expected to increase by a further 40% within the next two decades. Thus, today's water consumption is radically changed, and has reached levels that we can scarcely arrest, let alone comprehend. During the past 50 years, there have been more than 450 hostile water-related disputes, and on 37 occasions, countries have resorted to violent action (Galgano 2013). This problem is made more difficult because of shared water basins. At least 261 of the world's major rivers are shared, with 176 flowing through two countries, 48 through three, and 37 through four or more. These watersheds account for more than 45% of the land surface, they account for 60% of the our freshwater supply, and they supply nearly 40% of the world's population with water for domestic use, agriculture, hydropower, and other significant purposes (Butts 2011). As many as 80 countries—about two-fifths of the world's population—already suffer serious water deficits.

3.3 Population

Population growth is perhaps the core dynamic at the heart of most environmental trends. The time it takes to add a billion people to the global population is shrinking. For example, it took 130 years for world population to grow from one to two billion, but it only took about 14 years to jump from six to seven billion. Given the problems associated with failed states, the challenge of population growth is particularly problematical in the developing world in which more than 90% of the added billions will live. Here, sheer numbers do not translate into political power, especially when most of the added billions will be living in poverty—in states that are vulnerable to exposure to climate change and are essentially unable to adapt. The relationship linking population levels, the resource base, and the environment–climate nexus is complex. Adaptation is an essential function and governmental policies and institutions can spell the difference between a highly degraded environment and one that can provide for many more people.

4 Summary and Conclusions

Environmental issues are now a major factor in regional stability and an essential component of the global security landscape. Notwithstanding recent warfare in the Middle East and Afghanistan, a major issue confronting the U.N. and the West, since the end of the Cold War is conflict rooted in the conditions of the environment. This chapter developed the definitional framework that underpins environmental security and suggests that its associated trends will be central to the emergent national security landscape in the future. Environmental security is a highly interdisciplinary subject and the variety of backgrounds and motivations that researchers bring to the debate is beneficial because it encourages multiple disciplines to examine the dimensions of the environment–conflict nexus and offer greater solutions to policymakers. For example, the chapters in this book are replete with typologies of environmental conflict, with concepts based on variables such as, climate change, development, migration, and resource scarcity. This is possible because national security studies have moved decisively to include aspects of environmental security.

The evolution of the global strategic situation following the end of the Cold War has led to an acceptance of a broader definition of national security, which includes conflict and instability triggered or intensified by environmental factors, population pressure, and resource scarcity. Globalization has contributed to the environment–conflict nexus because it has eliminated much of the friction of distance and accelerated economic demands, leading to non-sustainable economic activity and environmental damage, which combined with population pressure and climate change has stressed many ecosystems beyond their capacity. It has also elevated the number of failed states incapable of keeping pace with the demands of environmental change; thus creating ungoverned spaces ripe for instability and conflict. As global population grows, economic demands may exceed the natural resource and economic base of many states, erode governmental legitimacy, destabilize regions, and promote intrastate conflict over increasingly scarce resources.

The scenario presented in this chapter and subsequent chapters in this book, clearly suggest that the future is not bright given our alteration of the natural environment and the weakening of government control in the developing world. Humans already consume 40% of the food and energy potentially available on land. While that percentage may be sustainable, it is unlikely that it could keep pace with expected increases in the world's population. Human use of some 80% of Earth's productivity is not compatible with the continued functioning of ecosystems and society as we know it—certainly a surge in environmentally triggered conflict is a plausible outcome.

Fortuitously, this gloomy prognosis is only a forecast based on contemporary trends. Like all projections it is rooted in present trends and the recent past; and does not inevitably reflect a precise image of the future. Hence, human society is not destined to enter a painful decline into environmentally triggered chaos. Clearly, there are scientific, technical, and economical solutions that are feasible to reduce the level of environmental stress and diminish potential conflict. Contemporary dynamics are replete with non-sustainable practices, along with ineffective governance that if corrected, could abate most environmental instability. However, these possibilities are confronted by extensive social, political, and institutional barriers. Unquestionably, our national security calculus demands the evaluation and consideration of environmental threats to security. The principle challenge to the U.S. security apparatus is that it is normally in the business of reacting to crises, and not traditionally equipped or staffed to plan for and deal with the non-traditional threats and challenges such as those presented by environmental security threats.

Literature Cited

Barnett TMP (2004) The Pentagon's new map. G. P. Putnam and Sons, New York
Bennett O (ed) (1991) Greenwar, environment and conflict. The Panos Institute, Washington, DC
Burke MB, Miguel E, Satyanath S, Dykema JA, Lobell DB (2009) Warming increases the risk of war in Africa. Proc Natl Acad Sci 106(49):508–537
Butts KH (2011) Environmental security: a growing force in regional stability. In: Galgano FA, Palka EJ (eds) Modern military geography. Routledge, New York, pp 54–64
Department of Defense (DoD) (2005) The national defense strategy of the United States of America. U.S. Government Printing Office, Washington, DC
Department of Defense (DoD) (2014) Quadrennial defense review. U.S. Department of Defense, U.S. Government Printing Office, Washington, D.C 88 p. http://www.defense.gov/pubs/2014_Quadrennial_Defense_Review.pdf. Accessed 6 May 2017
Drapeau MD, Mignone BK (2007) Culture, conflict, and … climate? Science 316(5831):1564
Dulian A (2004) Rwandan genocide. Int Aff 50(4):40–44
Femina F, Werrell CE (2012) Update: climate and security 101: why the U.S. National security establishment takes climate change seriously. The Center for Climate and Security, Briefer no. 23, 26 February 2014. https://climateandsecurity.org/reports/#BRIEFERS. Accessed 12 Jan 2017
Femina F, Werrell CE (2014) Climate change, drought, and social unrest. The Center for Climate and Security, Briefer no 11. https://climateandsecurity.org/reports/#BRIEFERS. Accessed 12 Jan 2017
Floyd R (2014) Analyst, theory, and security: a new framework for understanding environmental security. In: Floyd R, Matthew RA (eds) Environmental security approaches and issues. Routledge, New York, pp 21–35
Foster GD (2001) Environmental security: the search for strategic legitimacy. Armed Forces Soc 27(3):373–395
Galgano FA (2007) A geographic analysis of ungoverned spaces. The Pennsylvania geographer, 44(2): 67–90
Galgano FA (2013) Disasters and conflict: the Ogaden war of 1977. PA Geograph 51(2):66–90
Gleick PH (ed) (2012) The world's water, volume 7, the biennial report on freshwater resources. Island Press, Washington, DC
Harnish RL (2009) Environmental security information for the secretary. U.S. Department of State, Washington, DC
Hendrix C, Salehyan I (2012) Climate shocks and political violence: beyond scarcity, beyond Africa, Research brief–April 2012. Climate change and African political stability. The Robert Strauss Center, Austin
Homer–Dixon TF (1991) On the threshold: environmental changes as causes of acute conflict. Int Secur 2:76–116
Hsiang SM, Meng KC, Cane MA (2011) Civil conflicts are associated with global climate change. Nature 476:438–441
Intergovernmental Panel on Climate Change (IPCC) (2007) Impacts, adaptation, and vulnerability. Contribution of working group II to the fourth assessment report of the Intergovernmental Panel on Climate Change. Cambridge University Press, Cambridge, UK https://www.ipcc.ch/pdf/assessment-report/ar4/wg2/ar4_wg2_full_report.pdf. Accessed 8 May 2017
Intergovernmental Panel on Climate Change (IPCC) (2012) In: Field CB, Barros V, Stocker TF, Qin D, Dokken DJ, Ebi KL, Mastrandrea MD, Mach KJ, Platter GK, Allen SK, Tignor M, Midgley PM (eds) Managing risks of extreme events and disasters to advance climate change adaptation. Special report of working groups I and II of the Intergovernmental Panel on Climate Change. Cambridge University Press, Cambridge http://ipcc-wg2.gov/SREX/. Accessed 21 July 2013

Intergovernmental Panel on Climate Change (IPCC) (2014) Climate change 2014: impacts, adaptation, and vulnerability. Summary for policy makers. Working group II contribution to the fifth assessment report of the Intergovernmental Panel on Climate Change. Cambridge University Press, Cambridge, UK http://www.ipcc.ch/report/ar5/wg2/. Accessed 8 Nov 2016

Kaplan RD (2000) The coming anarchy: shattering the dreams of the post Cold War. Random House, New York

Kaufmann D, Kraay A, Mastruzzi M (2010) The worldwide governance indicators. Methodologies and analytical issues. The World Bank Development and Research Group. http://www.govindicators.org. Accessed 15 Feb 2017

Klare MT (2001) The new geography of conflict. Foreign Aff 80(3):49–61

Knickerbocker B (2007) Could global warming cause war? The Christian science monitor. Boston, MA, 19 April 2007, 2

Krakowka A (2011) The environment and regional security: a framework for analysis. In: Galgano FA, Palka EJ (eds) Modern military geography. Routledge, New York, pp 65–78

Maas A, Carius A, Wittich A (2014) From conflict to cooperation? Environmental cooperation as a tool for peace building. In: Floyd R, Matthew RA (eds) Environmental security approaches and issues. Routledge, New York, pp 102–120

Mutunga C, Zulu E, De Souza R (2012) Population dynamics, climate change, and sustainable development in Africa. Population Action International and African Institute for Development Policy, Washington, DC http://pai.org/wp-content/uploads/2012/09/Sustainable-Development-in-Africa.pdf. Accessed 12 Jan 2017

National Security Council (NSC) (1991) National security strategy of the United States. U.S. Government Printing Office, Washington, DC

Obama BH (2010) National security strategy. The White House, Government Printing Office, Washington, DC http://nssarchive.us/national-security-strategy-2010/. Accessed 3 Oct 2017

Percival V, Homer-Dixon (1995) Environmental scarcity and violent conflict: the case of Rwanda, Occasional Paper. Project on Environment, Population and Security, American Association for the Advancement of Science and University of Toronto, Washington, DC http://www.library.utoronto.ca/pcs/eps/Rwanda/rwanda1.htm. Accessed: 28 Jan 2017

Sappenfield M (2007) Six places in the world where climate change could cause political turmoil. The Christian Science Monitor, Boston 6 December 2007, 14

Shah S, Landay JS (2010) Pakistan crisis raises fears of destabilization. The Philadelphia Inquirer, 15 August 2010, A2

Smith D, Vivekananda J (2007) A climate of conflict. International Alert, London http://www.international-alert.org. Accessed 28 Dec 2016

Smith D, Vivekananda J (2009) Climate change, conflict, and fragility. International Alert, London http://www.initiativefor.peace.eu. Accessed 20 Feb 2017

Solow AR (2011) Climate for conflict. Nature 476:406–407

United Nations (U.N.) (1987) Our common future. Report of the world commission on environment and development, The Brundtland report. Oxford University Press, New York http://www.un.org/documents/ga/res/42/ares42-187.htm/. Accessed 23 Jan 2017

United Nations (U.N.) (2009) Water in a changing world, United Nations world water development report 3: http://publishing.unesco.org/. Accessed May 2012

Wagner CG (2005) The struggle for environmental security. The Futurist 39(6):14–15

White House (1988) National security strategy of the United States. Government Printing Office, Washington, DC http://www.state.gov/. Accessed 9 Nov 2017

White House (1991) National security strategy of the United States. U.S. Government Printing Office, Washington, DC

White House (1997) National security strategy of the United States. Government Printing Office, Washington, DC http://www.state.gov/. Accessed 9 Nov 2017

White House, 2010. National security strategy of the United States. Washington, D.C.: The White House. http://www.whitehouse.gov/sites/default/files/rss_viewer/national_security_strategy.pdf. (last accessed 2 August 2013)

Worldwide Governance Indicators (WGI) (2016) The worldwide governance indicators (WGI) dataset. The World Bank. http://info.worldbank.org/governance/wgi/index.aspx#home. Accessed 28 Dec 2016

Yohe G, Malone E, Brenkert A, Schlesinger M, Meij H, Xing X (2006) Global distributions of vulnerability to climate change. Integr Assess J 6(3):35–44

States at Risk: The Environment–Conflict Model

Francis A. Galgano

Abstract Linkages among global environmental problems and related economic, and demographic challenges have now emerged as one basis for interpreting conflict and security. This perspective has considerably refocused the lens by which we view the environment as a variable in the national security calculus. The Fourth Assessment Report of the IPCC suggested that institutions and governments, especially in the developing world, will have a great deal of trouble adapting to the strain of climate change. Although various environmental security models identify interrelationships of critical variables, they also underscore the principal weakness of these models, which is a lack of predictive capacity. This chapter presents a Vulnerability and Risk Index (VRI) to identify states at risk to environmentally trigged conflict. The VRI is a composite indicator derived from the geometric mean of five dimensions; and ranks 173 states from the most to the least vulnerable. VRI results suggest that the problem of risk is not limited to, but clearly highly concentrated in the developing world. The data indicate that 23 states are most vulnerable to violent conflict and 81 states are vulnerable to violent conflict—this represents 60% of the global population. The integration of information in an index, such as the VRI offers a useful quantitative indicator by which to assess risk related to exposure to climate change, vulnerability, and adaptive capacity.

Keywords Adaptation · Climate change · Climate vulnerability model · Composite indicator · Conflict · Developing world · Environmental degradation · Environmental security

F. A. Galgano (✉)
Department of Geography and the Environment, Villanova University, Villanova, PA, USA
e-mail: francis.galgano@villanova.edu

F. Galgano (ed.), *The Environment-Conflict Nexus*, Advances in Military Geosciences, https://doi.org/10.1007/978-3-319-90975-2_2

1 Introduction

During the past two decades there has been a dramatic shift in how we perceive the contemporary national security landscape. Leaders of governmental organizations and non–governmental agencies, as well as the scientific community have increasingly come to accept that the adverse effects of climate change and other environmental factors have exposed many vulnerable societies to instability and potentially, violent conflict. This altered perception of linkages among global environmental problems and related economic, and demographic challenges has now emerged as one basis for interpreting conflict and security (Femina and Werrell 2012). Indeed, the Intergovernmental Panel on Climate Change (IPCC) has devoted a great deal of effort to evaluating the vulnerability of human populations resulting from exposure to the adverse effects of climate change (IPCC 2007, 2012, 2014). In fact, the Fourth Assessment Report of the IPCC (2007), examines the issue of impacts, adaptation, and vulnerability and suggests that institutions and governments, especially in the developing world, will have a great deal of trouble adapting to the strain of climate change. Adaptation will be hindered by a lack of capacity and the hardest hit by these problems will be people living in poverty and within failing, unstable states.

This perspective has considerably refocused the lens by which we view the environment as a variable in the national security calculus. As population and economic demands escalate, and the adverse effects of climate change become more apparent, collectively these problems may disrupt vulnerable populations to the extent that they erode governmental legitimacy, thus making these states more vulnerable to conflict (Smith and Vivekananda 2009). Given these circumstances, it is plausible that we will witness a surge in three modes of conflict: internecine conflict driven by resources shortages, environmental stress, and demographic trends; civil war prompted by governmental collapse and economic factors; and limited–scale interstate conflicts.

Clearly, identifying states at risk to instability and violence from environmental causes involves an extensive and complex array of security issues, particularly if we define it very broadly. Furthermore, delineating factors that contribute to environmentally–triggered instability is an inexact method involving risk analysis based on complicated linkages between human and natural processes. Therefore, it is helpful to establish a framework to delineate the various factors that are operating in a place and from which cogent analyses can be made. Nevertheless, researchers and institutions have concluded that developing states are more vulnerable because they suffer from several persistent environmental and human variables such as environmental degradation; reduced agricultural production; economic decline; poor governance; population growth and displacement; and civil disruption (IPCC 2014).

Although environmentally triggered conflict can manifest itself in a variety of ways, and result from a wide diversity of contributing variables, Fig. 1 illustrates a common cycle in the environmental security paradigm (Homer-Dixon 1999). The cycle is initiated within a vulnerable society, which has insufficient ability to adapt. That population is exposed to adverse environmental change, or some environmental

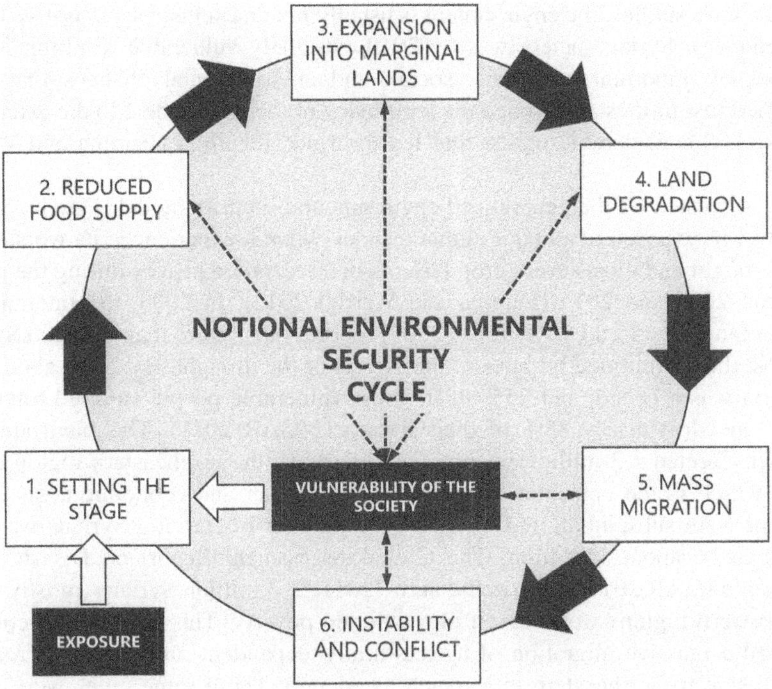

Fig. 1 A generic environmental security model depicting a common sequence of events. The civil war in Syria is an excellent example of this cycle (Femina and Werrell 2012)

shock such as an extensive drought or natural disaster. Thus, the stage is set by a combination of overpopulation, economic decline, social stratification, latent ethnic conflict, and a weak government. In stage two, exposure combines with extant environmental and societal issues to significantly reduce the food supply. This reduced agricultural output and food insecurity typically compels people to expand farming into marginal lands (Stage 3). The expansion into marginal lands, such as hillsides and marshlands, is essentially an unsustainable solution that leads to further environmental degradation from problems like deforestation, excessive soil erosion, soil exhaustion and desertification, and ultimately a complete failure of the agricultural system (Stage 4). The collapse of the agricultural system precipitates widespread farm abandonment and mass migration, usually to an urban environment or a nearby region that is perceived to offer essential resources (Stage 5). Mass migration typically overburdens the government and infrastructure, and places strains on ethnic fracture lines, typically leading to civil or internecine conflict.

The ongoing conflict in Syria is a useful example of the environment-conflict cycle described in Fig. 1. Clearly, however, Syria's current civil unrest is a consequence of its brutal and repressive government and near economic collapse; and it was also fostered by the general wave of political uprisings in the Middle East during the so–called Arab Spring (Welch 2015). Thus, like any number of environmental security case studies, the environment is usually not an exclusively causative factor. Nevertheless, Syrian society was by 2010 extremely vulnerable resulting from a number of important economic, social, and environmental changes that overwhelmed institutions and eroded the legitimacy of the government to the extent that it provoked considerable unrest that led to armed rebellion (Femina and Werrell 2012).

As suggested by the generalized environmental security model (Fig. 1), Syria's society was exposed to a major climatic shock when it experienced its worst long–term drought and most severe crop failures in its recorded history during the period between 2006 and 2011 (Femina and Werrell 2012). In 2009, the International Federation of the Red Cross (IFRC 2009) reported that more than 800,000 Syrians had lost their livelihood because of the effects of the drought. By 2011, about 60% of Syria was affected, and 75% of its most vulnerable people suffered total crop failure and lost nearly 85% of their livestock (GAR 2011). This environmental disaster affected 1.3 million people, particularly in the northeastern region of the state, which is the epicenter of the war. The human and economic costs of the drought were substantial; and by 2010, the number of starving Syrians was estimated to be about 1 million. The Global Assessment Report on Disaster Risk Reduction (GAR 2011) estimated that, by 2011, 2–3 million Syrians (mostly in the northeastern region) were driven into extreme poverty. This agricultural collapse initiated a massive migration of agriculturally–dependent rural families from the countryside to northeastern cities such as Aleppo. These same cities were at the same time trying to cope with the influx of Iraqi refugees (Femina and Werrell 2012). This massive influx of displaced people strained the adaptive capacity of an inadequate and failing government, strained ethnic and political tensions, and essentially delegitimatized the Syrian government. It is not a stretch to conclude that environmental conditions, and their effect on the population, enabled conditions that triggered the civil war (Welch 2015).

Although the model (Fig. 1) identifies the interrelationships of the critical variables and sequence of events that affect environmental security, it also underscores the principal weakness of the model, which is a lack of predictive capacity. That is, we have no overarching sense of which environmental scenarios will lead to conflict. Thus, the critical problem is defining a threshold at which a state may descend into violent conflict rather than adapt. Moreover, this chapter acknowledges that a conflict related to environmental factors cannot be predicted accurately. Rather it offers a theoretical framework and quantitative index by which we can identify states at risk to violent conflict given a set of critical environmental and human variables. Hence, it is an objective of this chapter to present a reasonable and plausible vulnerability index from which we may examine potential environmental security implications in the most vulnerable states and regions. The index is a simple and intuitive composite indicator of vulnerability based on the fewest number of critical variables and quantifies the vulnerability of states based on an index that ranges from 0.00 to 4.00.

2 The Environmental Conflict Vulnerability Model

A purpose of this chapter is to present an environmental security model that serves as a theoretical framework for a composite indicator that quantitatively identifies countries vulnerable to environmentally–triggered conflict. It is informed by previous models proposed by Homer-Dixon (1991, 1994, 1999), Twose (1991), Winfield and Morris (1994), Butts (1994), David (1996), Dalby (2002), Smith and Vivekananda (2007) Galgano and Krakowka (2010), and Floyd (2014); as well as IPCC (2001, 2007, 2014) assessments of impacts, adaptation, and vulnerability to climate change.

Homer-Dixon (1991) presented a Conflict Causality Model (Fig. 2), which is perhaps the seminal theoretical framework that explains the complex environment–conflict nexus by offering a causal–path analysis. As is the case with models, it is designed to simplify and bring order to the many complex and interrelated variables that contribute to this nexus. The Conflict Causality Model (Fig. 2) suggests that the link between the environment and conflict begins with the effects of human activity on the environment, which is essentially a function of two dimensions: population and human activity per capita (Homer-Dixon 1991). Resource availability, society, and governance influence human activity. The model (Fig. 2) suggests that human activity combines with ecosystem vulnerability to engender social effects that may ultimately trigger violent conflict. It is important to note that this is not a deterministic model, rather feedback loops suggest that governance, institutions, and elements of society may intercede and prevent conflict (Homer-Dixon 1991).

The introduction of the Conflict Causality Model (Fig. 2) is important because it establishes a theoretical basis for the essential dimensions that contribute to the environmental security paradigm and serves as a point of departure for models that followed. The elegance of the Conflict Causality Model (Fig. 2) is that it defines a set of causal linkages and identifies important relationships among the most

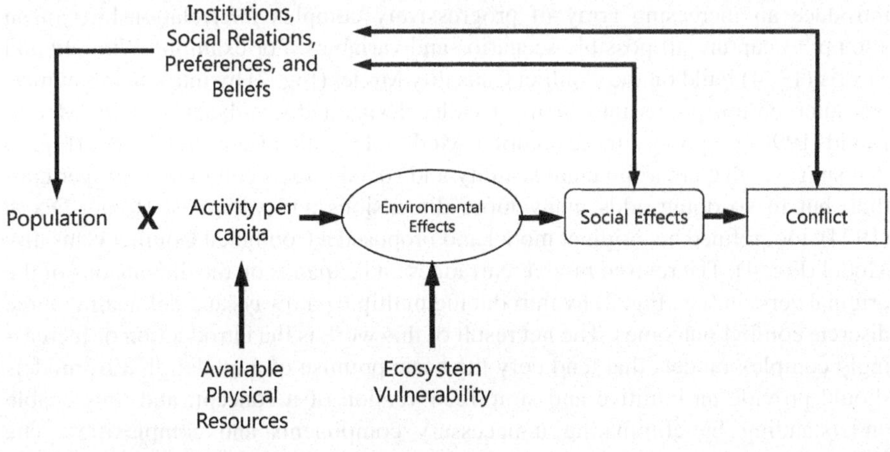

Fig. 2 Homer-Dixon's (1991) Conflict Causality Model

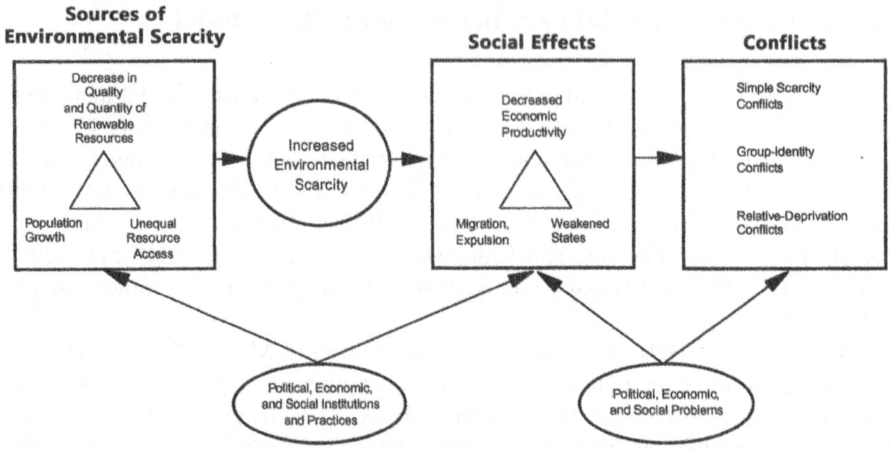

Fig. 3 David's (1996) Modified Conflict Causality Model

persistent and important human dimensions: i.e., human activity on the landscape, population, governance and society, and economic decline. Furthermore, it explains how human dimensions interact with critical elements of the environment, such as resource shortages (e.g., water, food, fuel, and forests), soil depletion and loss of arable land (i.e., agricultural production and food insecurity), ecosystem degradation, and environmental change (e.g., climate change, drought, pollution, etc.). The model acknowledges that the causal connections in the environmental security process are complex, and factors such as social beliefs, the fabric of society, and religion—by way of example—are difficult to assess and integrate into a global–scale model. Likewise, the myriad of localized environmental and societal issues are equally difficult to assess and integrate at larger spatial scales (Homer-Dixon 1991).

Consequently, since the introduction of the Conflict Causality Model, others have been proposed that expanded the number of dimensions and variables and introduce an increasing array of progressively complex interrelationships in an attempt to capture all possible scenarios and variables. For example, Winfield and Morris (1994) build on the Conflict Causality Model (Fig. 2) by introducing numerous intervention points into their so–called Expanded Conflict Causality Model. David (1996) proposes a more complex Modified Conflict Causality Model (Fig. 3) that suggests that environmental scarcity and social effects combine to trigger conflict, but in so doing adds many more dimensions and variables. Homer-Dixon (1994), too, refined his original model and proposed a Combined Conflict Causality Model (Fig. 4). The revised model rearranges and expands on the dimensions of the original version (i.e., Fig. 2) by introducing multiple pathways and delineating three discrete conflict outcomes. The net result of this work is the introduction of increasingly complex models that tend defy the basic premise of a model; that is, models should provide an intuitive and simplified version of a concept, and thus enable understanding by eliminating unnecessary components and complexities. The Homer-Dixon (1991, 1994) models and others that follow are nevertheless helpful

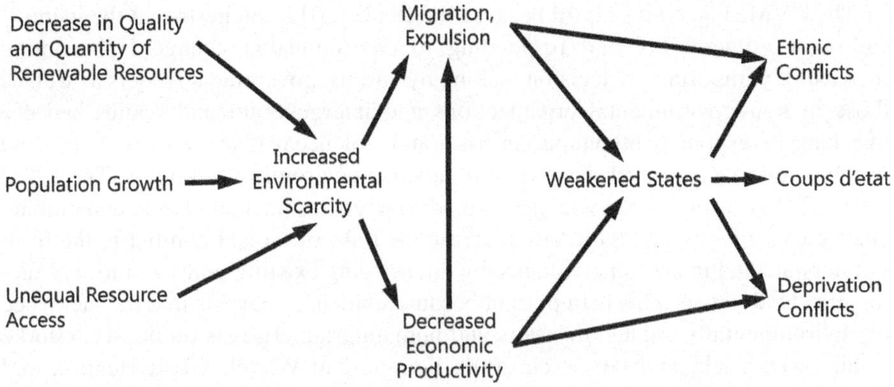

Fig. 4 Homer-Dixon's (1994) Combined Conflict Causality Model

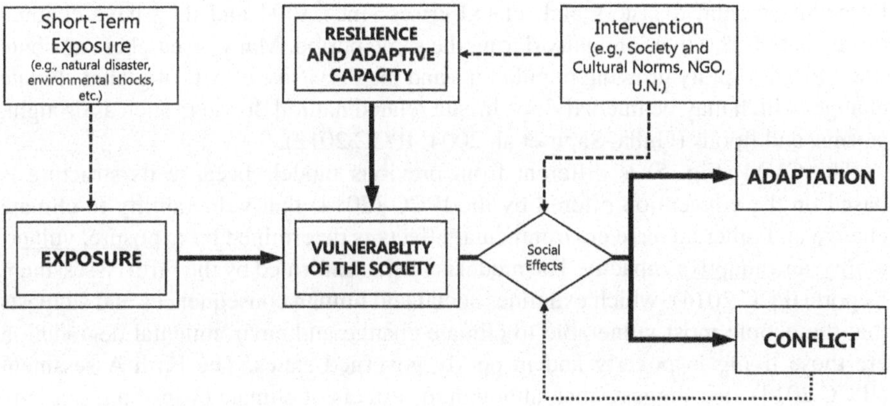

Fig. 5 The Conflict Vulnerability Model

because they delineate the persistent dimensions and relationships in the environment–conflict nexus: i.e., population and population dynamics; societal effects; governance; economics; human activity; resources; and environmental change.

This chapter proposes a much simplified Conflict Vulnerability Model (CVM) (Fig. 5). The CVM (Fig. 5) attempts to recognize and identify the relationships between people, society, and the environment that make selected places vulnerable to violent conflict. Thus, it provides a theoretical framework for a quantitative index. The CVM (Fig. 5) is not prescriptive or deterministic; more precisely it recognizes that vulnerability is just that; and it may be overcome by governmental or non–governmental institutions or factors of society that may provide the adaptive capacity to overcome vulnerability. Hence, in the case of the model's assessment of vulnerability, violent conflict is only one possible outcome; however, the model nevertheless provides a framework to identify states that are more vulnerable and where violent conflict is a plausible outcome.

The CVM (Fig. 5) is helpful because the IPCC (2012) suggests that the identification of vulnerability and responding to environmental change is becoming increasingly important in decision making by various government leaders as well as those in non–governmental organizations and intergovernmental bodies because they have to respond to humanitarian crises and violence in places adversely affected by climate–driven natural disasters and adverse environmental effects. The IPCC (2012, 2014) further acknowledges that adverse environmental effects and climate change can directly and indirectly increase the risks of violent conflict in the form of interstate and internecine violence by intensifying existing politico–military factors that lead to war. This is important because evidence suggests that the incidence of environmentally–related disasters and humanitarian crises is on the rise (Burke et al. 2009; Diehl and Gleditsch 2001; Femina and Werrell 2014; Hendrix and Salehyan 2012; Homer-Dixon 1999; Hsiang et al. 2011; Smith and Vivekananda 2009; Solow 2011; Yohe et al. 2006), and the number of people affected is increasing as well (IPCC 2012). One in twenty–five people worldwide have been affected by environmental disasters and related crises since 1994 and they have claimed more than 58,000 lives annually during the same period. Many researchers attribute this spike to rapidly growing populations and the pervasive effects of global climate change, which may be intensifying climate related natural disasters such as drought, storms, and floods (Guha–Sapir et al. 2004; IPCC 2012).

The CVM (Fig. 5) is different from previous models because its structure is based on the convention offered by the IPCC (2014) that vulnerability to climate change and other adverse environmental effects is determined by exposure, vulnerability, and adaptive capacity. The model is further informed by the Fifth Assessment Report (IPCC 2014), which examines social and human consequences and suggests that the people most vulnerable to climate change and environmental degradation are those living in poverty and in poorly governed states. The Fifth Assessment (IPCC 2014) also suggests that, although the effects of climate change are spatially variable and not linked to national boundaries, specific states are adversely and disproportionally affected by climate; and environmental change is making them individually more vulnerable and potentially predisposed to extreme political instability and violent conflict (Smith and Vivekananda 2009).

Thus, the goal is to develop a simplified model that logically links exposure, vulnerability, and adaptive capacity because they offer a universal lens by which to examine complex linkages among climate, society, and governance, and they provide a logical relational framework by which we may consider the central and persistent dimensions of the environment–conflict nexus. Exposure (Fig. 5) subjects societies to the adverse effects of climate and environmental change, thus negatively affecting the stability and security of people; and further degrading existing environmental conditions. Adaptation, which is essentially a function of governance and economic strength, defines the capacity of a society to cope and respond in an organized way to overcome the adverse effects of exposure and instability. Finally, vulnerability is a measure of existing social and environmental conditions in a state. Thus, the CVM (Fig. 5) suggests that a highly sensitive society with little ability to adapt to the adverse effects of exposure is likely to be more vulnerable to extreme

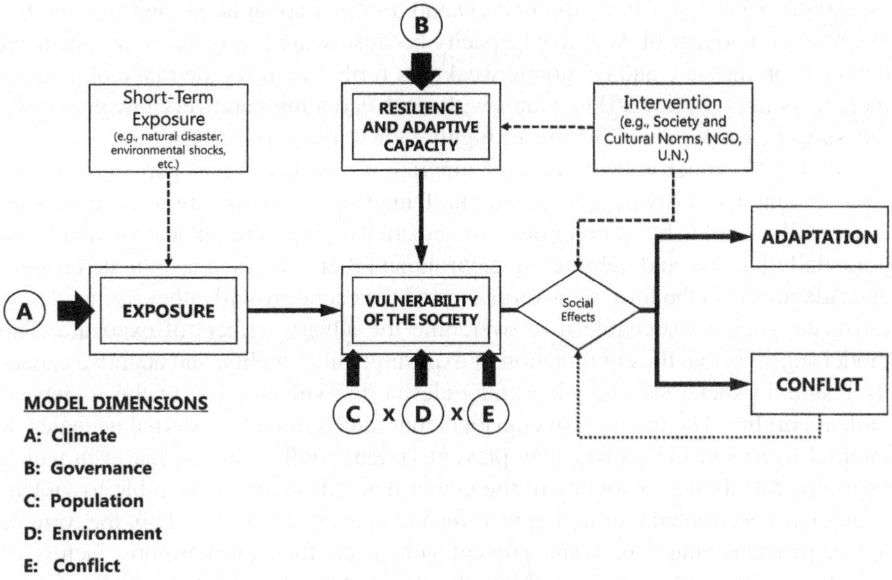

Fig. 6 The Conflict Vulnerability Model with key dimensions

political instability and violent conflict if there is ineffective or no intervention from governmental institutions, the stabilizing effects of cultural norms, or aid from an external organization such as the U.N. (Dalby 2002).

To provide a theoretical framework for a quantitative index, the CVM is further modified by the interaction of five discrete dimensions that define the model's operational parameters. The model (Fig. 6) suggests that the link between the environment and conflict begins with Vulnerability, which is the product of three dimensions: Population (i.e., C, Fig. 6), Environment (i.e., D, Fig. 6), and Conflict (i.e., E, Fig. 6). Vulnerability is the central parameter in the model because it defines existing social conditions as well as the availability of resources and the environmental conditions within a state, which delineates its predisposition to instability and violence (Dalby 2002).

In the model (Fig. 6), Exposure is defined by Climate (i.e., A, Fig. 6), which may include episodic shocks, such as a natural disaster. Climate and its adverse effects account for the persistent and long–term exposure of a society that may magnify extant conditions of vulnerability. Exposure, especially in the case of episodic events, may also overwhelm the ability of governmental institutions, the economy, and society to adapt or intervene (IPCC 2014).

A single dimension, Governance (i.e., B, Fig. 6), defines Adaptive Capacity. Governance as well as economic strength underlie the problem of vulnerability because poor governance and political instability make it difficult to overcome exposure. Many state functions are of central importance to adaptation. States with effective governments are better situated to mitigate the effects of exposure and balance sensitively because they typically have many robust institutions to provide essential services and infrastructure (David 1996). Economic strength is another essential component of Adaptive Capacity because scarcities increase the financial demands on the state and economic weakness further narrows the range of government responses (Solow 2011). States with weak economies and very low *per capita* measures of wealth are indicative of highly vulnerable populations.

The CVM model is centered on Vulnerability as the critical parameter in the environmental security conflict paradigm. Dimensions of population, environment, and conflict establish the conditions of sensitivity. Exposure subjects a society to potentially adverse and debilitating environmental effects, such as climate change; and Adaptation in the form of governance and the economy will either further destabilize the society, or enable it to overcome the adverse effects of exposure. The model suggests that the combination of exposure, vulnerability, and adaptive capacity results in social effects, which may elevate the vulnerability of the society to violent conflict. The model also considers that intervention by external agencies or internal factors of the society may prevent violent conflict. In a scenario in which exposure destabilizes a society to the extent that it becomes vulnerable to violent conflict, a governmental or non–governmental agency may intercede in the form of aid or peacekeeping forces and prevent violent conflict. Furthermore, factors of society and cultural norms may alter the threshold to instability and violence as well.

The CVM (Fig. 6) is an intuitive and simplified model based on contemporary literature that suggests that vulnerability to environmentally triggered instability and conflict is a function of exposure, vulnerability, and adaptive capacity. These widely accepted factors make the model universal and essentially scale–independent since the five related dimensions offer the ability to introduce relevant variables enabling it to be used at different geographic scales. Thus, the CVM (Fig. 6) performs three functions. First, it provided a simple and intuitive framework by which we may examine factors and relationships that expose a society. Second, the model attempts to reduce the complex variables and relationships in the environment–conflict nexus by defining five key dimensions. Finally, the CVM also considers that intervention can take place to mitigate the effects of exposure, compensate for poor governance, and prevent a place from devolving into conflict. The model also recognizes that the causal connections in the environmental security process are complex and factors such as the fabric of society are difficult to measure and quantitatively integrate into a model, especially at a global scale. Likewise, localized environmental conditions are complex and equally difficult to assess and integrate into the model.

3 Composite Indicator of States at Risk

A composite indicator is developed when individual variables are compiled into a single–value index based on a multidimensional concept that provides a theoretical framework—in this case, the environment–conflict nexus (Fig. 6). Composite indices are useful tools for summarizing multidimensional concepts such as human development, environmental sustainability, economic productivity, and industrial competitiveness, which cannot be effectively represented by many discrete variables (Saltelli 2007). Composite indicators have been used for many years to index economic data, such as the Consumer Price Index (BLS 2017), which is calculated by collecting the prices of representative consumer items over discrete time periods. The United Nations Development Program (UNDP) has for many years published the Human Development Index, which is a composite index of achievement in three key dimensions of human development (UNDP 2015).

An index is compiled through a mathematical combination of a set of related variables. The index should be based on a theoretical framework that enables variables to be selected, weighted, and combined in a manner reflecting the structure of the concept (Munda 2005). Composite indices can be highly effective because they enable—in the context of this chapter—comparisons of country performance, or provide a ranking of countries, and facilitate the identification of complex trends across many separate variables, which may also be adapted to various geographic scales. Furthermore, when developed and evaluated at regular temporal intervals, an index may also facilitate the identification of the magnitude and direction of important trends (Saltelli 2007). Hence, composite indicators are particularly effective because the single–value index is less complicated and easier to interpret than a series of variables; and they reduce the visible size of a set of variables without eliminating the underlying base of information (OECD 2008). Nevertheless, the strength or weakness of a composite index is strongly linked to the selection and quality of its underlying variables. Although the choice of variables is guided by their relevance and relationship to a theoretical framework, the selection process and decisions on weighting can be somewhat subjective.

Efforts to develop environmentally related composite indicators such as vulnerability, risk, or development indices are not new. Indexing heterogeneous data into a composite indicator such as the Vulnerability and Risk Index (VRI), developed for this analysis, is different because it produces a single–value measure of vulnerability that takes into account human and environmental factors. Thus, the VRI simultaneously considers anthropogenic and environmental variables in a theoretical framework, whereas previous indices have focused social vulnerabilities or environmental factors, not necessarily a combination of the two; and certainly not in relation to an environmental security framework. For example, IPCC (1991), Pernetta (1990), and Downing (1992) developed indices based on climate change and sea level rise. Atkins et al. (1998) and UNEP (1998) examined ecosystem vulnerability indices based on economic susceptibility and human impacts on the environment, while Pantin (1997) indexed the impact of natural disasters. Nixon and Qiu (2005) developed a multi–variable storm severity index. Yohe et al. (2006) developed a global index that illustrated vulnerability to climate change.

There have been attempts to develop composite indices that combine anthropogenic and environmental variables; however, few have been focused explicitly on environmental security. Kaly et al. (1999) and Kaly et al. (2004) assembled an environmental risk exposure index that integrated social, economic, and environmental data and rank–ordered small island countries by defining their relative environmental vulnerabilities. This index, however, focused on human effects on the environment rather than environmental security as such. The University of Notre Dame Global Adaptation Initiative (ND-GAIN) has instituted a project called the ND-GAIN Country Index (ND–GAIN 2017) that summarizes a state's vulnerability to climate change and other global challenges in relation to its adaptive capacity and resilience. This composite index is compiled by merging two dimensions (i.e., vulnerability and readiness) that encompass 36 vulnerability–related variables and nine variables that contribute to the measure of readiness. However, this is not an environmental security index, per se, rather it aspires to enable organizations in the business and the public sectors to prioritize investments and facilitate efficient responses to global challenges rather than assess vulnerability in a national security sense (ND–GAIN 2017).

Mutunga et al. (2012) developed an index to demonstrate the implications of population growth and climate change for sustainable development in sub–Saharan Africa. In their assessment, they developed a series of indices that depicted the combined influence of agricultural production, population, water scarcity, and climate change. The core of their analysis was a composite indicator, the Vulnerability–Resistance Indicators Model, that combined 17 physical, social, and economic indicators to measure the resilience of African states to climate change impact (Mutunga et al. 2012). This index was designed to indicate a state's ability to recover from the effects of climate change, given its sensitivity to existing human and environmental conditions (e.g., food security, human health, water scarcity) and adaptive capacity (e.g., economic and governmental resources, and the environment).

Perhaps the best example of an environmental security composite index was developed by the London based non–profit group International Alert (i.e., Smith and Vivekananda 2007), which examined the nexus of climate change and failing states. Smith and Vivekananda (2007) suggest that 46 states are at a high risk of violent conflict by the end of this century; and another 56 face a high risk of political instability. Smith and Vivekananda (2007) based their index three broad criteria: governance and economic stability, vulnerability to climate change, and an operational presence of peacekeeping forces. Their results are very compelling because their index implies that environmental security is adversely affecting some 52% of the world's countries. However, their index does not offer a single–value indicator that differentiates among states at risk; rather it creates three broad groups of states: (1) states facing a high risk of violent conflict; (2) states facing a high risk of political instability; and (3) other states. Their findings are given in a choropleth map that coalesces groups of states into these broad categories and does not further differentiate or rank order among the countries, thus implying that vulnerability is equal in each broad category. Furthermore, their "other states" category implies that little or no vulnerability is associated with these countries, thus masking potential levels of vulnerability of about 48% of the world's states.

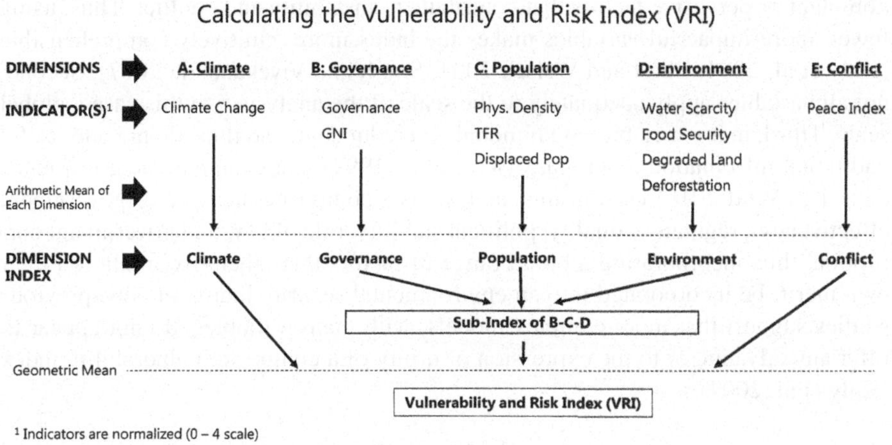

Fig. 7 Procedure for assembling the Vulnerability and Risk Index (VRI)

3.1 Theoretical Framework for the VRI

The objective of this chapter is to, using a single–value global scale environmental security index, identify and rank states at risk to instability and violent conflict. The VRI is predicated on the principle that human and environmental systems are interdependent, and that their interaction within the multidimensional CVM (Fig. 6) will translate into a level of vulnerability. Figure 6 illustrates a conceptual model that identifies the operational components of environment–conflict nexus, and thus provides a theoretical framework for the creation of this index. The VRI is a simple and intuitive composite indicator of vulnerability derived from the geometric mean of five dimensions (Fig. 7). The dimensions (or sub–indices) are compiled from normalized, heterogeneous variables that define the complex interplay of human and environmental factors in the ECM (Fig. 6).

Perhaps the most important decision in formulating the VRI is the selection of the variables incorporated in the index. The CVM (Fig. 6) suggests that human and environmental systems are dependent on one another. The exposure of a society to factors such as climate change and other environmental variables, combined with human dynamics affect the vulnerability of the state. Clearly, there are numerous variables that potentially contribute to environmental security; however, the objective of the VRI is to identify and rank–order states at risk based on the fewest number of critical indicators—in this case, at a global scale. In the case of this index, it was developed using variables related to the persistent dimensions and relationships in the environment–conflict nexus: i.e., population and population dynamics; societal effects; governance; economics; human activity; resources; and environmental change (Homer-Dixon 1991).

Although it may be argued that a larger number of variables may be required in a vulnerability assessment, research suggests that typically, there are only a few dominant or pervasive factors that contribute to instability and conflict. Thus, using fewer, more impactful variables makes the index more intuitively comprehensible (Kaly et al. 2004; Ebert and Welsch 2004; Smith and Vivekananda 2007). Second, not all variables apply adequately to the scale of the analysis—in this case a global scale. Third, many variables are bimodal and redundant, and thus, do not add useful additional information to an index (Kaly et al. 1999). For example, the governance data (i.e., WGI 2016) used in this analysis is a composite measure of government effectiveness, regulatory quality, political stability, rule of law, and government corruption, thus incorporating a broad range of factors that otherwise might, on their own merit, be incorporated into an environmental security index. Finally, previous studies suggest that more complex models, with many variables, do not appear to offer any advantages to the expression or utility of a composite vulnerability index (Kaly et al. 2004).

3.2 *Index Variables*

The VRI is the geometric mean of five human/environmental dimensions, which are linked to the CVM (i.e., A, B, C, D, and E, Fig. 6). The dimensions, or sub–indices, are further composed of 11 individual variables (or indicators) that characterize components of exposure, vulnerability, and adaptation (Table 1). The variables were selected because each is essentially global and thus applicable over the scale of the study. Moreover, they are well defined, easy to understand, and the data are sufficiently robust to support their inclusion. Finally, they enable a global comparison because the resultant index can sufficiently differentiate among states and also demonstrate regional disparities. Thus, the VRI defines the vulnerability of each state by reducing complex, multidimensional, and heterogeneous variables into a single–value index using an easy to understand scale from zero (the most vulnerable state) to four (the least vulnerable state).

The Climate sub–index is defined by a single variable that expresses a state's exposure to the adverse effects of climate change. In many regions, climate change is affecting human and natural systems and the consequences are becoming more clear (IPCC 2014). The variable used to define this sub–index is exposure to climate change from the Notre Dame Global Adaptation Initiative (ND–GAIN 2017). This variable defines a state's exposure to the negative effects of climate change; and is essential because exposure reduces resiliency, increases vulnerability, lowers the threshold for violent conflict and exposure to the significant effects of climate change.

The Governance sub–index is quantified using two variables, Governance and Gross National Income (GNI). Although the effects of climate change are not confined by state boundaries, the effects of environmental security are not randomly and fairly distributed because adaptive capacity is a function of the strength of the state and financial resources of the population. Thus, the effects of environmental security are defined by state boundaries and governance is a key indicator.

Table 1 VRI dimensions and indicators

Dimension (Sub–index)	Variable (Indicator)
A: Climate	Vulnerability to Climate Change (value)[a]
B: Governance	Governance (value)[b]
	Gross National Income (GNI) Per Capita (dollars/per capita/year)[c]
C: Population	Physiologic Density of Population (inhabitants/km[b] arable land)[d]
	Total Fertility Rate (value)[e]
	Displaced Population (total displaced population)[f]
D: Environment	Total Internal Renewable Water Resources per capita (m^3/per capita)[g]
	Food Supply (percent of population undernourished)[h]
	Population Living on Degraded Land (percent)[i]
	Deforestation (hectares/year)[j]
E: Conflict	Domestic and Internal Conflict (value)[k]

[a]ND Gain Data Set, http://index.gain.org/ranking/vulnerability
[b]World Bank Governance Indicators, http://info.worldbank.org/governance/wgi/index.aspx#home
[c]World Bank, World DataBank, http://datacatalog.worldbank.org/
[d]Food and Agriculture Organization of the United Nations, http://www.fao.org/nr/water/aquastat/data/query/index.html?lang=en (total population / km[b] arable land)
[e]United Nations, UN Data, http://data.un.org/Data.aspx?d=PopDiv&f=variableID%3A54
[f]Internal Displacement Monitoring Center, http://www.internal-displacement.org/database/
[g]Food and Agriculture Organization of the United Nations, http://www.fao.org/nr/water/aquastat/data/query/index.html?lang=en
[h]Food and Agriculture Organization of the United Nations, http://www.fao.org/nr/water/aquastat/data/query/index.html?lang=en
[i]Food and Agriculture Organization of the United Nations http://www.fao.org/nr/water/aquastat/data/query/index.html?lang=en
[j]Food and Agriculture Organization of the United Nations, http://www.fao.org/nr/water/aquastat/data/query/index.html?lang=en
[k]Global Peace Index, http://economicsandpeace.org/research/

Furthermore, the IPCC (2014) suggests that the people most affected by the problematic effects of climate change are those living in poverty in which fragile governance and weak economies erode their livelihood and security. World Bank Governance Indicator data (WGI 2016) are compiled by monitoring six key metrics of governance: i.e., voice and accountability of the government; political stability; government effectiveness; regulatory quality; rule of law; and control of corruption. The data are comprehensive, and have been compiled essentially on a yearly basis between 1996 and 2015. GNI *per capita* (World Bank 2017) is the value of the total domestic and foreign economic output claimed by the residents of a state and is expressed in U.S. Dollars *per capita*. GNI is an effective indicator because it has proved to be closely correlated with quality of life and the ability of a population to overcome the adverse effects of the environmental conditions and climate change (World Bank 2017).

Three sub–indices quantify Vulnerability and factors of Population are of central importance. In this sub–index, population dynamics are a function of three variables that express population density, growth, and migration. Physiologic density (i.e., the number of people per unit area of arable land) is an insightful variable because it is an excellent indicator of stress in societies that are highly dependent on the land for basic subsistence. To express population growth Total Fertility Rate (TFR) was selected as the second populations variable. TFR is a synthetic value indicating the average number of children born to a women of child bearing age over her lifetime. It is a better indicator of fertility and population growth than the crude birth rate because it is independent of the age structure of the population (U.N. 2017). The third variable is the number of displaced people in a state. These data are available from the Internal Displacement Monitoring Center, Global Internal Displacement Database (IDMC 2017).

The Environment sub–index was calculated using four indicators: i.e., water scarcity (FAO 2017), percent of population undernourished (FAO 2017), percent of people living on degraded land (FAO 2017), and deforestation (FAO 2017). Water scarcity is a leading indicator of environmental stress because of the depletion of fresh water resources (Mathews 1989). Water scarcity is a variable expressed as renewable water resources (i.e., $meters^3$) *per capita*. Food is a necessity of life and food security is an essential element to the basic security of a population. This variable was derived by dividing the total number of people undernourished in each state by total population to obtain a useful expression of food insecurity (FAO 2017). This variable is advantageous because in the developing world, major crops such as wheat, rice, and maize are increasingly vulnerable to failure. Related to food and water shortages is the number of people living on degraded land, which is a strong indicator of climate change, agricultural failure, and demographic pressures. Similarly, the amount of deforested land is an important expression of environmental security because it is strongly correlated with the expansion of urban areas and farming into marginal agricultural lands.

The sub–index for Conflict is quantified by one variable that measures the presence of extant conflict with a state. A common characteristic of failed states is the persistence of internal conflict, which may be exacerbated by factors of environmental security (Femina and Werrell 2014). The presence of violence is ranked in 163 states, representing 99.7% of the world population, by the Global Peace Index (GPI 2016) using a 10–year longitudinal assessment of 23 qualitative and quantitative variables. The GPI is based on the level of safety and security in a society; the extent of domestic and international conflict; and the degree of state militarization.

3.3 The Vulnerability and Risk Index

The overriding principle of the VRI is to present a single–value composite index of vulnerability that is intuitively comprehensible: i.e., the variables are sufficient to identify and underscore differences between states that are most or least vulnerable;

and second, to develop an easy to calculate index by employing a user–friendly interface in Microsoft Excel. The first step (Fig. 7) is the reduction of the 11 individual variables to a common scale (i.e., "0" to "4") by normalizing their data (Equation 1), which enables their combination by averaging. Normalization is required because the indicators are heterogeneous and expressed in disparate units and scales. Normalization eliminates heterogeneity and compensates for highly skewed data thus producing a common scale between all variables by which comparisons and calculations may be accomplished (Ebert and Welsch 2004). Equation 1 is a so–called "min–max" form of normalization and was selected for its simplicity, but more importantly, it widens the range of the variables lying within small intervals, thus accentuating their influence on the index (OECD 2008).

$$\text{Equation1}: z_i = \left(x_i - \min(x)\right) \cdot (b-a) / \left(\max(x) - \min(x)\right) \tag{1}$$

Following normalization, several indicators required an inversion to ensure that their normalized scale was consistent with the design of the index: i.e., zero equates to most vulnerable and four equates to the least vulnerable state. For example, a state with a large percentage of population living on degraded land is problematical, but its normalized value would be closer to "4.0" thus incorrectly suggesting that it is less vulnerable. To correct this situation, normalized scores for several variables (i.e., physiologic density, total fertility rate, displaced population, percent population living on degraded land, and deforestation) were inverted using Eq. 2.

$$z_i = \left(n_i \bullet (-1)\right) + 4 \tag{2}$$

Normalized variables are compiled into sub–indices using their arithmetic mean (Fig. 7). Finally, the VRI was assembled by computing the geometric mean of the dimension sub–indices using Eq. 3. The model (Fig. 7) indicates that sensitivity is the product of its dimensions. In this case the population, environment, and conflict sub–indices are combined by multiplication into a single "super" sub–index. The geometric mean indicates the central tendency of, or typical value for, a set of numbers defined by the *nth* root of the product of *n* numbers and is appropriate for the creation of this type of index because it produces a more meaningful expression of central tendency when comparing data with multiple properties, so that no particular dimension dominates the mean (U.N. 2015).

$$VRI = \left(I_{Climate} \bullet I_{Governance} \bullet \left(I_{Population} \bullet I_{Environment} \bullet I_{Conflict}\right)\right)^{1/3} \tag{3}$$

Two variables are weighted. Weights can have a significant effect on country rankings in a composite index, and yet the decision to weight is essentially reduced to a value judgment (Ebert and Welsch 2004). Many composite indices rely on the equal weighting of variables, which implies that they are valued the same in the construction of index. This method can be useful because it acknowledges the absence of a clear understanding of causal relationships in the model. However, if variables are grouped into dimensions, as they are in the VRI, then equal weighting may imply an unequal weighting of the dimensions because those with a greater number of variables will have a greater impact on the final composite index, which may lead to an unbalanced structure and final value (OECD 2008). Thus, in constructing the index, two variables were weighted, each by a factor of two: i.e., governance and climate.

In calculating the VRI, the composite index had to account for missing information and a method for replacing missing data with substituted values (i.e., data imputation) was considered and rejected. Although one of the selection criteria for VRI variables was the completeness of the data set, there were consistent information gaps, particularly for smaller states and state–like entities (e.g., Monaco, Vatican City, West Bank and Gaza) and small island states (e.g., Kiribati, Palau, Grenada). Missing data can clearly hinder the robustness of a composite index. Data can be missing in a random or non–random manner and there are essentially three accepted methods for dealing with this problem: case deletion, single imputation, and multiple imputation (OECD 2008). The missing values in the VRI variables are not missing at random, that is the data are missing systematically because they are associated with a consistently defined subset of states and in all data sources. Values that are not missing at random are difficult to substitute because methods to impute are based on the assumption that the data are missing because of some random mechanism. Thus, the VRI used the case deletion method, which simply omits the data from the analysis. This is considered acceptable because the overall sample size of states is sufficiently large to support the overall robustness of the final index (OECD 2008). Thus, the final VRI excludes 29 states from the final composite index (Table 2). The rational for using case deletion involved two primary criteria. First, if data were missing from the single variable dimensions (i.e., Climate and Conflict), the state was removed from the VRI because their inclusion would produce an incomplete and biased index value. Second, if a state was missing more than one of the variables from the multivariable dimensions (i.e., Governance, Population, and Environment), it was deleted because its inclusion would create a systematic bias (OECD 2008). In all cases, missing data were not given a value of zero so as not to influence the value of a mean score.

3.4 Strengths and Weaknesses of the VRI

As is the case with composite indicators, the VRI is defined by a number of methodological strengths and weaknesses that have to be understood to ensure its proper application and to place results in an appropriate context. First, the VRI is comprehensive in scope, and as a composite indicator, reduces the number of variables without ignoring essential base of information. It is effective for expressing vulnerability in the environmental security model because the single–value index is less complicated and easier to interpret than a series of discrete variables. Second, the VRI combines critical human and environmental variables and thus, offers a more comprehensive evaluation of environmental security. Third, the VRI more completely differentiates between states and identifies their level of vulnerability using

Table 2 States removed from the VRI related to missing values[a]

Country/Territory	WB Code	Climate	Governance	Population	Environment	Conflict
American Samoa	ASM	X		X	X	
Andorra	ADO	X				
Anguilla	AIA	X		X	X	
Aruba	ABW	X		X	X	
Bermuda	BMU	X				
Cayman Islands	CYM	X		X		
French Guiana	GUF	X		X		
Grenada	GRD	X				
Guam	GUM	X				
Kiribati	KIR	X				
Macao Sar, China	MAC	X		X	X	
Marshall Islands	MHL	X				
Martinique	MTQ	X		X	X	
Micronesia, Fed. Sts.	FSM	X			X	
Monaco	MCO	X		X		
Nauru	NRU	X		X	X	
Palau	PLW	X			X	X
Réunion	REU	X		X	X	X
Samoa	WSM				X	X
San Marino	SMR				X	
Seychelles	SYC				X	X
St. Kitts And Nevis	KNA					X
St. Lucia	LCA					X
Suriname	SUR					X
Tonga	TON					X
Tuvalu	TUV				X	X
Vanuatu	VUT					X
Vatican City	VAT	X	X	X	X	X
West Bank And Gaza	WBG	X				

[a]The "X" indicates a Dimension with missing value(s)

an intuitive, single–value index. Finally, the VRI is user friendly and its simplicity enables its adaptation for the creation of vulnerability indices at various scales and permits periodic updates to identify trends.

Perhaps the most important weakness of the VRI is that it is dependent on state–level data and these data may mask important regional variations within a country. However, given the global scope of this analysis, this shortcoming was considered to be acceptable. Second, common to all indices, the strength or weakness of the VRI is clearly associated with the selection of the underlying variables. Third, although the choice of variables is guided by their relevance and relationship to the multidimensional concept under study, decisions on variable weighting can be somewhat subjective. Fourth, because of missing data values, 29 states are excluded from the VRI. The lack of data that generated these deletions is a common problem across multiple variables, sources, and data, but more importantly, given that the index provides a comprehensive vulnerability assessment for 173 other states, this shortcoming was deemed to be necessary. Finally, the index does not take into account short–term effects that are part of the CVM (Fig. 6), such as an abrupt natural disaster event, or the intervention of an NGO or introduction of peacekeeping forces, by way of example.

4 Results and Discussion

The interaction of factors and variables in the, the CVM (i.e., Fig. 6) is predicated on the analytical structure offered by the IPCC (2007) that risk to environmentally–triggered instability depends on factors of exposure, vulnerability, and adaptive capacity. The VRI is based on 11 critical variables related to five dimensions that contribute to the model's dynamics relationships: i.e., climate, governance, population, environment, and conflict (Figs. 6 and 7). VRI results are illustrated in the map given in Fig. 8. The map assigns one of four colors to each of the 173 states that comprise the index. These states include nearly 87% of the global population. A list of the most and least vulnerable states are given in Table 3. VRI data are broken down as follows:

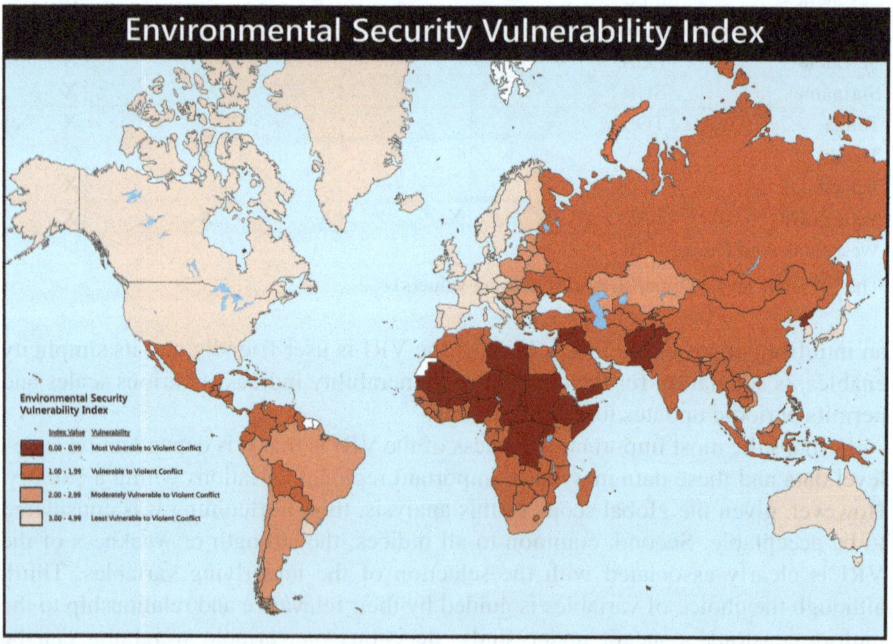

Fig. 8 Choropleth map illustrating the results of the VRI. The data suggest that risk and vulnerability are highly concentrated in the developing world

Table 3 Results of the VRI indicating the states most and least at risk to conflict triggered by climate exposure

States most vulnerable to violent conflict		States least vulnerable to violent conflict	
State	VRI Value	State	VRI Value
Somalia	0.00	Uruguay	3.01
Chad	0.05	Portugal	3.04
South Sudan	0.15	Italy	3.16
Sudan	0.26	Belgium	3.16
Central African Republic	0.27	Czech Republic	3.17
Afghanistan	0.28	Japan	3.26
St. Vincent And The Grenadines	0.33	Spain	3.31
Yemen, Rep.	0.33	Netherlands	3.32
Eritrea	0.46	United States	3.36
Burundi	0.50	France	3.37
Congo, Dem. Rep.	0.51	Austria	3.45
Pakistan	0.54	Germany	3.50
Iraq	0.70	United Kingdom	3.53
Ethiopia	0.76	New Zealand	3.61
Mauritania	0.77	Denmark	3.61
Nigeria	0.78	Finland	3.62
Mali	0.79	Ireland	3.62
Libya	0.80	Sweden	3.64
Korea, Dem. Rep.	0.81	Luxembourg	3.69
Uganda	0.82	Australia	3.77
Niger	0.89	Canada	3.83
Burkina Faso	0.93	Iceland	3.93
Guinea-Bissau	0.95	Norway	3.99
		Switzerland	3.99
		Greenland	4.00

- The 23 dark red states (13.2% of the sample size) have values between 0.00 and 0.99, and represent the states that are <u>most</u> vulnerable to violent conflict.
- The next grouping of states have values between 1.00 and 1.99 and represent the states that are vulnerable to violent conflict. These 81 states (46.8% of the sample size) are shaded in medium red.
- States framed in light red have index values between 2.00 and 2.99. These 44 states (25.4% of the sample size) are <u>moderately</u> vulnerable to violent conflict.
- The last grouping of states, those identified by pink shading, are states that are <u>least</u> vulnerable to violent conflict. These 25 states (14.4% of the sample size) have values between 3.00 and 4.00.
- States not considered by the index are colored white.

The interpretation of the index and the assignment of colors in Fig. 8 are informed by similar assessments given by the IPCC (2001, 2014) and Smith and Vivekananda (2007). The results, as expressed by the spatial pattern of colors (Fig. 8) suggest that developed and developing countries alike could be at risk to instability triggered by

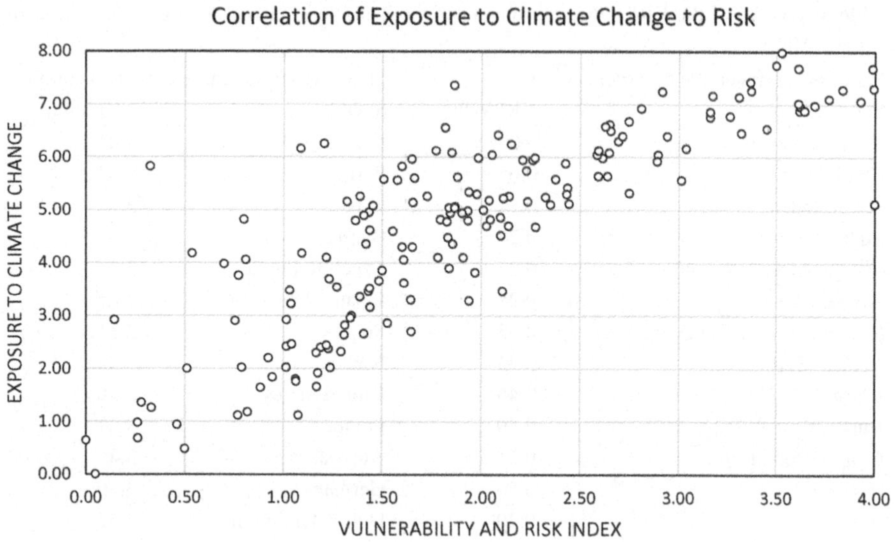

Fig. 9 Correlation of exposure data to the Vulnerability and Risk Index

climate and other related environmental factors. Nevertheless, the data clearly suggest that the problem of vulnerability and risk is greatly magnified in the developing world. All of the most vulnerable states (i.e., values between 0.00 and 0.99) are located there; and furthermore, 77 of the 81 states identified in the second grouping—states vulnerable to violent conflict—are concentrated in the developing world as well. Without question exposure to environmental effects negatively influences the stability and security of people. As suggested in Fig. 9, the climate exposure variable is highly correlated to the VRI. Especially in the developing world, within which large segments of the population are highly dependent on the land for their subsistence, states are more vulnerable to instability because they suffer from four related effects: (1) reduced agricultural production; (2) economic decline; (3) population displacement; and (4) civil disruption (Homer-Dixon 1991). These problems define the underlying vulnerability of the society and undermine adaptive capacity (IPCC 2014).

The VRI clearly suggests that developing states, within which governance is chronically ineffective, are in general more problematical in the environmental security model. These states are negatively affected by burgeoning population and population dynamics; social stratification and latent ethic conflict; poor governance; weak economies; a lack of vital resources; food insecurity; and environmental degradation. These factors collectivity make them vulnerable and diminish their ability to adapt to climate change and other environmental shocks. Even with rapid advances in adaptive capacity or intervention by the stabilizing effects of a society or external agency, these states will have trouble keeping pace with exposure to climate and environmental impacts. Although the effects of climate change are not confined by state boundaries, the effects of the environment–conflict nexus are not

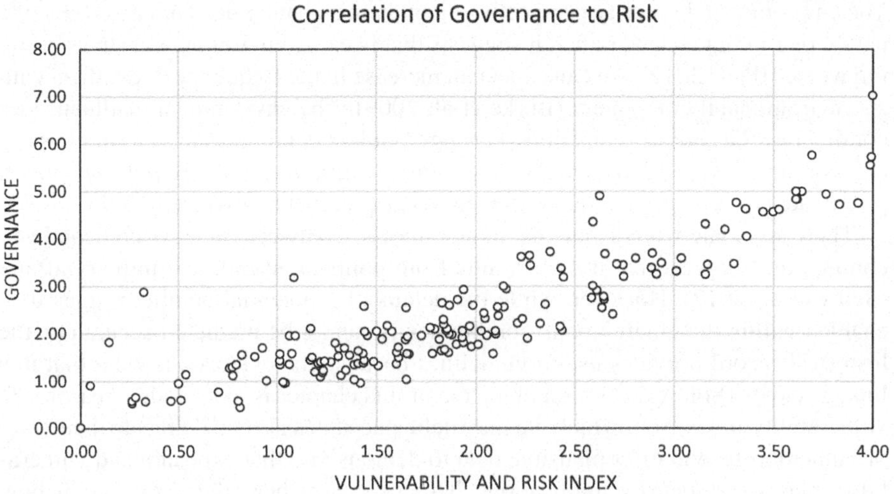

Fig. 10 Correlation of governance data to the Vulnerability and Risk Index

randomly and fairly distributed because adaptive capacity is a function of the strength of the state and the financial resources of the population. Thus, the effects of environmental security are defined by state boundaries and governance is a key indicator. The IPCC (2014) suggests that the people most affected by the problematic effects of climate change are those living in poverty in which fragile governance and weak economies erode their livelihood and security. The data given in Fig. 10 indicate the very strong correlation between governance and the VRI.

The issue of ineffective governance in the developing world is critical. The number of failing states is growing, and adaptive capacity and stability is tied strongly to governance (Smith and Vivekananda 2009). Failing states are problematical because they have large areas that are outside of effective government control and are thus affected negatively and severely by humanitarian disasters, environmental stress, and internecine conflict. Hence, states in the developing world are more susceptible to several specific modes of conflict: internecine conflict driven by resources shortages, environmental stress, and demographic trends; civil war prompted by governmental collapse and economic factors; and limited–scale interstate conflicts. Each mode of conflict has serious repercussions for the security interests of the developed world because conflicts with an environmental component, coupled with divisive ethnic dimensions, such as those observed in Syria, have increased pressure on the West and U.N. to commit resources to stability efforts (Femina and Werrell 2014).

This chapter narrowed, considerably, the scope of its research by focusing on how environmental and climate change affects conflict, rather than a much broader interpretation of environmental security—nevertheless, the issue of the environment–conflict nexus is very broad. Climate and environmental exposure may influence conflicts as dissimilar as interstate war, internal civil conflict, insurgencies, terrorism, resource disputes, terrorism, or trade disagreements (Homer-Dixon 1999).

The environmental security perspective given in this chapter does not assert that the nature of conflict is new; rather, it suggests that because environmental stress growing worse (IPCC 2012), we can expect an increase in the frequency of conflicts with an environmental component (Burke et al. 2009). The environment's influence on conflict may be causative and directly trigger violent conflict, such as a simple scarcity conflict (David 1996). Or, the environment's role may only be a powerful, yet proximate cause as seen in the Syrian civil war (Femina and Werrell 2012).

There are some who dispute the interconnection between the environmental and conflict; and suggest that it only results from political, social, and military affairs (Selby et al. 2017). However, while the details of a potential conflict triggered or enabled within the environment–conflict nexus cannot be predicted accurately, the historical record provides useful guidelines because the evidence is clear that this linkage exists (Solow 2011). An objective of this chapter is to provide a reasonable vulnerability index from which we may delineate states at risk, delineate their level of vulnerability, and offer plausible data to demonstrate that exposure and vulnerability may have security implications. This of course begs the questions of how might environmental factors trigger conflict? Case studies suggest that the most plausible scenario is when environmental degradation reduces food supplies and focuses pressure along long–standing ethnic fault lines, which could generate internecine conflict (Fig. 1). Exposure to climate and environmental effects ultimately initiates a gradual impoverishment of societies in poorly governed states that could exacerbate class and ethnic cleavages, destabilize governments and spawn insurgencies or intra–state war. At the international level, some scholars propose that environmental change may shift the balance of power within regions, creating instabilities over critical resources that could lead to war (Dalby 2002). Swelling populations living on highly stressed land have the potential for the mass migration of environmental refugees that can spill across borders with destabilizing effects on domestic order and international stability. Finally, the prevalence of transboundary watersheds in already unstable areas is a recipe for interstate conflict as is the new problem of contention over ice–free sea lanes in the Arctic (Homer-Dixon 1994).

5 Summary and Conclusions

The security of people, states, and regions is being increasingly jeopardized by climate–related and environmental threats. The data generated in this analysis suggest that the risk is global; but, it is not equally distributed around the world—it appears to be a pervasive problem concentrated in the developing world. The environmental security paradigm is the most interdisciplinary of interdisciplinary topics, yet agreement on the subject does not exist among scientists or security specialists. Hence, the environment–conflict nexus is sometimes not seen as a security or as an environmental issue. Nevertheless, environmental security is a concern because, even without directly causing conflict, the adverse effects of the environment have the potential to destabilize governments, harm and displace populations, and trigger violent conflict. Consequently, the integration of information in an index, such as the VRI, offers a new lens by which to assess risk related to exposure, vulnerability, and adaptive capacity. This integration provides insights into the conditions within which adaptive capacity and resilience may or may not be able to impart what is required so that states may adapt in a timely manner. The VRI results suggest that the problem of risk is not limited to, but is clearly highly concentrated in the developing world. Furthermore, poorly governed states are projected to experience impacts of climate change more readily and may be overwhelmed. Finally, the VRI data, while substantially suggestive, are unquestionably reliant on the framing of the vulnerability index and the variables selected.

Literature Cited

Atkins JP, Mazzi S, Easter CD (1998) A commonwealth vulnerability index for developing countries. Commonwealth Secretariat, London

Bureau of Labor Statistics (BLS) (2017) Consumer price index. https://www.bls.gov/cpi/. Accessed 3 Jan 2017

Burke MB, Miguel E, Satyanath S, Dykema JA, Lobell DB (2009) Warming increases the risk of war in Africa. Proc Natl Acad Sci 106(49):508–537

Butts KH (1994) Environmental security: A DoD partnership for peace. Army War College Strategic Studies Institute, Carlisle Barracks

Dalby S (2002) Environmental security. University of Minnesota Press, Minneapolis

David WE (1996) Environmental scarcity as a cause of violent conflict. Master of Military Arts and Science thesis. U.S. Army Command and General Staff College, Fort Leavenworth

Diehl PF, Gleditsch NP (eds) (2001) Environmental conflict. Westview Press, Boulder

Downing TE (1992) Climate change and vulnerable places: global food security and country studies in Zimbabwe, Kenya, Senegal, and Chile. Research report 1, Environmental Change Unit, University of Oxford. http://www.ciesin.org/docs/004-042/004-042.html. Accessed 15 Feb 2017

Ebert U, Welsch H (2004) Meaningful environmental indices: a social choice approach. J Environ Econ Manag 47:270–283

Femina F, Werrell CE (2012) Update: climate and security 101: chy the U.S. national security establishment takes climate change seriously. The Center for Climate and Security, Briefer no. 23, 26 February 2014. https://climateandsecurity.org/reports/#BRIEFERS. Accessed 12 Jan 2017

Femina F, Werrell CE (2014) Climate change, drought, and social unrest. The Center for Climate and Security, Briefer no 11. https://climateandsecurity.org/reports/#BRIEFERS. Accessed 12 Jan 2017

Floyd R (2014) Analyst, theory, and security: A new framework for understanding environmental security. In: Floyd R, Matthew RA (eds) Environmental security approaches and issues. Routledge, New York, pp 21–35

Food and Agriculture Organization of the U.N. (FAO) (2017) Aquastat. http://www.fao.org/nr/water/aquastat/data/query/index.html?lang = en. Accessed 23 Dec 2016

Galgano FA, Krakowka A (2010) The environment's influence on regional stability and conflict. PA Geograph 48(2):11–40

Global Assessment Report on Disaster Risk Reduction (GAR) (2011) Revealing risk, redefining development. United Nations, International Strategy for Disaster Reduction. https://www.preventionweb.net/english/hyogo/gar/2011/en/home/download.html. Accessed 3 Feb 2017

Global Peace Index (GPI) (2016) Global peace index. Institute for Economics and Peace. https://www.economicsand peace.org. Accessed 2 Feb 2017

Guha–Sapir D, Hargitt D, Hoyois P (2004) Thirty years of natural disasters 1974–2003: the numbers. Presses University of Louvain, Louvain–la–Neuve

Hendrix C, Salehyan I (2012) Climate shocks and political violence: beyond scarcity, beyond Africa, Research brief–April 2012. Climate change and African political stability. The Robert Strauss Center, Austin

Homer-Dixon TF (1991) On the threshold: environmental changes as causes of acute conflict. Int Secur (2):76–116

Homer-Dixon TF (1994) Environmental scarcities and violent conflict: evidence from the cases. Int Secur (1):5–40

Homer-Dixon TF (1999) Environment, scarcity, and violence. Princeton University Press, Princeton

Hsiang SM, Meng KC, Cane MA (2011) Civil conflicts are associated with global climate change. Nature 476:438–441

Intergovernmental Panel on Climate Change (IPCC) (1991) The seven steps to the vulnerability assessment of coastal areas to sea–level rise: guidelines for case studies. Response Strategies Working Group; Steering Committee for IPCC Sea Level Rise Assessment – Coastal Zone Management. http://trovenlagovau/work/24258331?selectedversion=NBD9978734 Accssed 10 Dec 2016

Intergovernmental Panel on Climate Change (IPCC) (2001) Impacts, adaptation, and vulnerability—technical summary. Cambridge University Press, Cambridge

Intergovernmental Panel on Climate Change (IPCC) (2007) Impacts, adaptation, and vulnerability. Contribution of Working Group II to the Fourth Assessment Report of the Intergovernmental Panel on Climate Change. Cambridge, UK: Cambridge University Press. https://www.ipcc.ch/pdf/assessment-report/ar4/wg2/ar4_wg2_full_report.pdf. Accessed 8 May 2017

Intergovernmental Panel on Climate Change (IPCC) (2012) Managing risks of extreme events and disasters to advance climate change adaptation. Special Report of Working Groups I and II of the Intergovernmental Panel on Climate Change, Field CB, Barros V, Stocker TF, Qin D, Dokken DJ, Ebi KL, Mastrandrea MD, Mach KJ, Platter GK, Allen SK, Tignor M, and Midgley PM (eds). Cambridge University Press, Cambridge. http://ipcc-wg2.gov/SREX/. Accessed 21 July 2017

Intergovernmental Panel on Climate Change (IPCC) (2014) Climate change 2014: impacts, adaptation, and vulnerability. Summary for policy makers. Working Group II Contribution to the Fifth Assessment Report of the Intergovernmental Panel on Climate Change. Cambridge University Press, Cambridge. http://www.ipcc.ch/report/ar5/wg2/. Accessed 8 Nov 2016

Internal Displacement Monitoring Center (IDMC) (2017) Global Internal Displacement Database. http://www.internal-displacement.org/database/

International Federation of the Red Cross and Red Crescent (IFRC) (2009) Syria: drought. http://www.ifrc.org/docs/appeals/09/MDRSY001do.pdf. Accessed 3 Jan 2018

Kaly UL, Briguglio L, McLeod H, Schmall S, Pratt CR, Pal R (1999) Environmental vulnerability index (EVI) to summarize national environmental vulnerability profiles. SOPAC Technical Report 275

Kaly UL, Pratt C R, Mitchell J (2004) The demonstration environmental vulnerability index (EVI) 2004. SOPAC Technical Report 384

Mathews JT (1989) Redefining security. Foreign Aff 68(2):162–177

Munda G (2005) Multi-criteria decision analysis and sustainable development. In: Figueira J, Greco S, Ehrgott M (eds) Multiple-criteria decision analysis. Springer, New York

Mutunga C, Zulu E, De Souza R (2012) Population dynamics, climate change, and sustainable development in Africa. Population Action International and African Institute for Development Policy, Washington, DC http://pai.org/wp-content/uploads/2012/09/Sustainable-Development-in-Africa.pdf. Accessed 12 Jan 2017

Nixon WA, Qiu L (2005) Developing a storm severity index. J Transp Res Board 1911:143–148

Notre Dame Global Adaptation Initiative (ND–GAIN) (2017) ND–Gain country index. http://index.gain.org/ranking/vulnerability/exposure. Accessed 24 Jan 2017

Organization for Economic Development and Cooperation (OECD) (2008) Handbook for constructing composite indicators: methodology and users guide. Composite Indicators Research Group. https://composite-indicators.jrc.ec.europa.eu/?q=content/oecd-jrc-handbook. Accessed 11 Dec 2016

Pantin DA (1997) Alternative ecological vulnerability indices for developing countries with special reference to small island developing states. Report to the U.N. Department of Economic and Social Affairs

Pernetta JC (1990) Projected climate change and sea level rise: a relative impact rating for the countries in the Pacific Basin. In: Pernetta JC, Hughes PJ (eds) Implications of expected climate changes in the South Pacific Region: and overview. UNEP Regional Seas Report 1990. https://wedocs.unep.org/rest/bitstreams/13341/retrieve. Accessed 8 Jan 2017

Saltelli A (2007) Composite indicators between analysis and advocacy. Soc Indic Res 81:65–77

Selby J, Dahi OS, Fröhlich C, Hulme M (2017) Climate change and the Syrian war revisited. Polit Geogr 60:232–244

Smith D, Vivekananda J (2007) A climate of conflict. International Alert, London http://www.international-alert.org. Accessed 28 Dec 2016

Smith D, Vivekananda J (2009) Climate change, conflict, and fragility. International Alert, London http://www.initiativefor.peace.eu. Accessed 20 Feb 2017

Solow AR (2011) Climate for conflict. Nature 476:406–407

Twose N (1991) What is greenwar? In: Bennett O (ed) Greenwar: environment and conflict. The Panos Institute, Washington, DC

United Nations Environmental Program (UNEP) (1998) Human development report 1998. UNDP Report. Oxford, U.K.: Oxford University Press, 228 pp. http://hdr.undp.org/sites/default/files/reports/259/hdr_1998_en_complete_nostats.pdf (accessed 23 January 2017)

United Nations (U.N.) (2017) UN data, total fertility rate. http://data.un.org/Data.aspx?d=PopDiv&f=variableID%3A54. Accessed 3 Jan 2017

United Nations (U.N.) (2015) Human Development Report 2015. Technical Notes. http://hdr.undp.org/sites/default/files/hdr2015_technical_notes.pdf (accessed 1 February 2017)

United Nations Development Program (UNDP) (2015) Human development report, 2015. New York. United Nations Development Programme. http://hdr.undp.org/sites/default/files/2015_human_development_report.pdf. Accessed 1 Mar 2017

Welch C (2015) Climate change helped spark Syrian war, study says. National Geographic, 2 March 2015. https://news.nationalgeographic.com/news/2015/03/150302-syria-war-climate-change-drought/. Accessed 2 Feb 2018

Winfield JA, Morris ME (1994) Where environmental concerns and security strategies meet: green conflict in Asia and the Middle East. Rand, Santa Monica

World Bank (2017) World DataBank. Gross National Income. http://datacatalog.worldbank.org/. Accessed 2 Feb 2017

Worldwide Governance Indicators (WGI) (2016) The worldwide governance indicators (WGI) dataset. The World Bank. http://info.worldbank.org/governance/wgi/index.aspx#home. Accessed 28 Dec 2016

Yohe GE, Malone A, Brenkert M, Schlesinger H, Meij Y, Xing X (2006) Global distributions of vulnerability to climate change. Integr Assess J 6(3):35–44

Defining Climate Change: What to Expect in a Warmer World

Adam J. Kalkstein

Abstract Climate change is likely to have important security implications on a global scale, yet most discussions covering this topic fail to outline or define the specific environmental impacts that should be expected in a warmer world. At present, there is overwhelming scientific evidence that average surface temperatures across the planet are rising, with the best science currently suggesting most if not all of the observed warming is due to human activity. The expected impacts of climate change on the physical environment are wide-ranging and include stronger, but not necessarily more frequent hurricanes. Tornado activity is unlikely to change appreciably in the future, although sea level will continue to rise, possibly at a faster pace. Future flooding is likely to become worse, partially due to increasing atmospheric water vapor, while drought is also expected to become more common as a result of heightened evaporation. Severe heat waves, a particularly dangerous natural disaster for the mid-latitudes, are also expected to become more frequent and intense. A firm understanding of climate change and its most likely impacts are a vital component to any discussion relating climate change and the security environment.

Keywords Anthropogenic · Carbon dioxide · Climate change · Climate signal · Drought · Environmental security · Flood heat · Glaciation · Greenhouse effect · Greenhouse gas · Hurricane · Little ice age · Medieval warm period · Methane · Natural environment · Precipitation · Sea level rise · Security · Temperature · Thunderstorm · Tornado · Tropical cyclone · Typhoon · Water vapor

A. J. Kalkstein (✉)
Department of Geography and Environmental Engineering, Center for Languages, Cultures, and Regional Studies, United States Military Academy, West Point, NY, USA
e-mail: adam.kalkstein@usma.edu

F. Galgano (ed.), *The Environment-Conflict Nexus*, Advances in Military Geosciences, https://doi.org/10.1007/978-3-319-90975-2_3

1 Climate Change Background

Climate change is the most pressing environmental issue currently facing the planet, and it has the potential to have profound impacts on the environment–conflict nexus. Despite the obvious effects a changing climate might have on security, examining the relationship between the two has only emerged as an area of interest in recent years. Further, most discussions of climate change and security are nebulous; climate change is rarely defined, and its specific impacts on Earth processes are seldom outlined. To completely understand any potential relationship between climate change and environmental security, it is essential to first define what climate change is, discuss its causes, and most importantly, highlight its potential impacts on the physical environment.

Today, nearly all scientists agree the surface of the planet is warming, and evidence of this warming exists across numerous branches of physical science beyond Meteorology and Climatology. The vast majority of glaciers across the planet are receding, migration patterns of birds are changing, flowering dates of plants are becoming earlier, ocean temperatures are rising, sea ice across the planet is declining, and ocean levels are climbing (Walther et al. 2002; Root et al. 2003; Xu et al. 2009; Nerem et al. 2010; Huang et al. 2017a; NSIDC 2017). From Biology to Oceanography, Chemistry to Geology, virtually all scientific evidence points to the fact that temperatures on Earth are increasing.

Although there is little doubt average global temperatures are on the rise, the warming has not been uniform, and estimates of the magnitude of the warming vary somewhat. Data from The National Aeronautics and Space Administration (NASA) suggest global temperatures have increased by approximately 0.92 °C since 1880, and estimates from the National Oceanic and Atmospheric Administration (NOAA) are similar (NASA 2017; NOAA 2017a). Lower tropospheric air temperatures measured reliably by satellite since the late 1970's also show a modest warming of between 0.114 °C per decade and 0.174 °C per decade (Spencer et al. 2015; Mears and Wentz 2017). Finally, The Intergovernmental Panel on Climate Change (IPCC), a global consortium of leading climate scientists, estimates average temperatures across the planet have increased by approximately 0.85 °C from 1880 through 2012 (IPCC 2013).

While the general pattern of planetary warming is easy to discern, determining its causes is more complex. Temperatures across Earth can fluctuate naturally due to a variety of factors including orbital variations, changing solar output, volcanic eruptions, the movement of continents, changing ocean currents, and beyond. For example, humans played no role in warming Earth out of its most recent period of glaciation which ended around 12,000 years ago. During this time, temperatures were much lower than today, and nearly 30% of Earth's land was covered in ice. Likewise, human activity was not responsible for more recent but modest temperature changes such as the "Medieval Warm Period" or subsequent "Little Ice Age". Thus, separating any anthropogenic climate signal from natural fluctuations is particularly challenging.

Despite the known changes in Earth's climate that have occurred naturally, it is important to note that the current period of warming is occurring against the backdrop of rising greenhouse gas concentrations. These gases which include carbon dioxide, methane, and water vapor, are increasing due to human activity such as the burning of fossil fuels and continuing industrialization. The physical properties of greenhouse gases are well–established; while they generally allow the light of the sun to penetrate the atmosphere, they have the ability to absorb some of the outgoing heat from Earth's surface. Most scientists believe the increasing levels of these gases act to enhance Earth's natural "greenhouse effect," thus warming the planet. The scientific evidence supporting this theory of anthropogenic warming is robust, and current research suggests the majority of the present-day warming, if not all of it, can be traced to human activity. Although natural causes cannot be ruled out entirely for a portion of the warming, the IPCC (2013) recently concluded: "*It is extremely likely that human influence has been the dominant cause of the observed warming since the mid–20th century*." with "extremely likely" defined as 95–100% certainty. To summarize, there is little doubt temperatures across Earth are on the rise, with human activity being the most likely driving force.

2 Impacts on the Natural Environment

The largest source of scientific uncertainty surrounding climate change centers on the specific impacts it will have on the natural environment. Explicitly outlining these expected impacts, including highlighting the uncertainty associated with them, is essential prior to any discussion of climate change and environmental security.

2.1 Hurricanes and Tropical Cyclones

Tropical cyclones have historically been among the deadliest natural disasters and can have profound impacts on the security environment. These storms, also referred to as hurricanes and typhoons depending on location, require warm, deep water to develop, and bring with them a variety of hazards including storm surge, flooding, and extreme winds. In addition to the strength and size of the storm, the specific impacts of tropical cyclones on humans are dependent upon a variety of factors and include land surface cover, coastal geography, local infrastructure, warning systems, emergency planning, and beyond.

The relationship between climate change and tropical cyclones is surprisingly complex. Although it is well–established that ocean temperatures are rising resulting in additional "fuel" for tropical cyclones, warm water is only one of many factors necessary for these storms to strengthen. In fact, tropical cyclones require a highly specific set of environmental ingredients to be in place beyond warm water such as an unstable atmosphere, low vertical wind shear, high humidity, and favorable upper-level wind flow. Climate change is expected to impact each of these ingredients differently, somewhat muddying the future outlook for these systems.

Estimates for future tropical cyclone frequency across the planet vary considerably. While most studies conclude that global tropical cyclone activity will likely decrease in frequency, there is conflicting evidence in the North Atlantic Ocean, where storm frequency might be more closely related to water temperature than in the other ocean basins (Bengtsson et al. 1996; Mann and Emanuel 2006; Saunders and Lea 2008; Walsh et al. 2016). Across the rest of the tropics, the expected global decrease in tropical cyclone frequency is primarily a result of two factors. Although ocean temperatures will continue to rise, wind shear is expected to increase and the atmosphere will likely become more stable, both ingredients detrimental to tropical cyclone formation (Knutson et al. 2010; Kang and Elsner 2015).

The science regarding the impact of climate change on tropical cyclone intensity is more straightforward. To summarize, when atmospheric conditions are favorable for tropical cyclone development, the higher ocean temperatures will allow the storms that do form to become more intense (Webster et al. 2005; Oouchi et al. 2006). Scientists expect a general 2–11% increase in tropical cyclone intensity across the planet, and the number of particularly intense hurricanes (winds above 130 mph) in the Atlantic Basin is expected to double by 2100 (Bender et al. 2010). In addition to the changing intensity of future tropical cyclones, there is also evidence suggesting that the geographic extent of hurricane formation is expanding (Kossin et al. 2014), and the length of hurricane season has the potential to increase (Dwyer et al. 2015). Clearly, the relationship between climate change and tropical cyclones is complex, and although the number of storms might not change appreciably in the future, it is likely that the storms that do form will become stronger and potentially threaten new geographic locations.

2.2 Tornadoes and Severe Thunderstorms

As a result of its unique geography, more tornadoes are reported in the United States than any other country. Throughout the spring and early summer, when warm, moist air surging north from the Gulf of Mexico comes in contact with cold air from the North and dry air from the West, conditions are nearly ideal for severe weather outbreaks across the central United States. However, while the atmospheric conditions necessary for severe weather are well–understood, the potential impact of climate change on severe weather remains unclear.

The number of reported tornadoes across the United States has been increasing steadily for at least 60 years, although most scientists agree this is a result of better equipment, observation, and reporting rather than climate change (NOAA 2017b). As evidence, scientists point to the frequency of the strongest tornadoes, those that would have been reported in both the past and present. Among these particularly intense tornadoes, there has been very little change in frequency over time, with perhaps a slight decrease being observed. Only the weakest tornadoes, those most likely to have gone unreported in the past, have shown a marked increase over time, again providing evidence that the overall increase in tornado reports is related to a change in observation rather than climate.

Much like tropical cyclones, the formation of severe thunderstorms and tornadoes requires a specific set of atmospheric conditions to be present including moist, unstable air, wind shear, and contrasting air masses. Although climate change is expected to cause some of these ingredients to become more common in the future, others will become less frequent. More specifically, while instability and water vapor across the United States are likely to increase in the future, wind shear is likely to decrease (Trapp et al. 2007; Brooks 2013). As a result, most scientists agree that conditions favorable for thunderstorm formation might become more common in the future, but the reduction in wind shear will act to inhibit tornado formation somewhat, especially for the most intense tornadoes.

While the impact of climate change on severe weather and tornadoes is mixed, there is increasing evidence that the both timing and location of severe weather will change in the future. For example a larger percentage of overall tornadoes is occurring during concentrated tornado outbreaks (Elsner et al. 2015). Further, as surface temperatures increase, storm tracks are expected to move poleward, possibly altering the locations most at risk for severe weather (Bengtsson et al. 2006). In all, most scientists do not expect the average strength or frequency of tornadoes to change appreciably in the future, although there is some evidence that the location, timing, and frequency of severe weather outbreaks might fluctuate somewhat.

2.3 Sea Level Rise

Sea level has been rising since the last glacial maximum occurred approximately 20,000 years ago, often at rates much faster than today. At present, average global sea level is rising by roughly 3.4 mm per year, primarily due to thermal expansion, a result of rising ocean temperatures, and glacial melt (Beckley et al. 2017). Although the rate of sea level change has remained reasonably constant over the past few thousand years, there is at least some evidence suggesting that oceans are beginning to rise at a faster pace than has been observed in recent years, possibly a result of climate change (Rignot et al. 2011). However, not all scientists agree sea level rise is accelerating, and to date, there is no scientific consensus on the issue (Haigh et al. 2014).

While global ocean levels continue to increase, smaller–scale geographical features can act to exacerbate this increase at the local level. As one example, the melting of the Laurentide Ice Sheet, which in some places was over two miles thick, has resulted in a dramatic reduction of weight on the surface of the Earth over the past 10,000 years. Through a process called isostatic rebound, the Earth's crust has responded by rising in some locales but sinking in others. For example, in portions of Canada where the ice sheet was thickest, the elevation has risen by over 10 mm per year. However, in other places beyond the periphery of the historical ice sheet such as the mid-Atlantic region of the United States, the land has been sinking, albeit at a more modest rate (Sella et al. 2007). In these locations, decreasing elevation along the coast serves to exacerbate local sea level rise. Other factors contributing to local sea level change include the erosion of river deltas through a loss of sediment, largely a result of upstream dams and levees. This phenomenon is especially pronounced along the Mississippi River Delta, where scientists estimate Louisiana has lost 1883 square miles between 1932 and 2010, with a rate exceeding 16.5 square miles per year since 1985 (Couvillion et al. 2011). Finally, changing currents can also cause local variations in sea level. As one example, evidence suggests the Gulf Stream has slowed along the East Coast of the United States in recent years, likely a result of climate change. This has resulted in a local sea level rise several times faster than the global average (Sallenger et al. 2012; Ezer et al. 2013).

An additional and potentially catastrophic source of sea level rise centers on the possible collapse of the Western Antarctic Ice Sheet. This portion of Antarctica sits below sea level, making it more susceptible to melting than other areas of the continent. Recent studies have suggested that western Antarctica, which contains 10% of the ice across the continent, has already begun an "irreversible collapse" and will contribute 0.25 mm per year of sea level rise in the near future (Joughin et al. 2014; Scambos et al. 2017). As the melting accelerates, the contribution of the Western Antarctic Ice Sheet will increase to over 1.0 mm per year within the next couple hundred years.

2.4 Floods

The physical properties of the atmosphere dictate that at higher temperatures, more water can exist in its gaseous state. As a result, it is no surprise that water vapor has been increasing across the planet (Santer et al. 2007). Most scientists believe the rise in water vapor is at least partially responsible for an observed increase in extreme precipitation events in many locales, including the northern and eastern portions of the United States (Groisman et al. 2004). Further, scientists estimate that the most extreme precipitation events will more than double with every degree increase in global temperature (Trenberth 2011; Shiu et al. 2012). Land-use changes such as an increase in urbanization and corresponding impermeable surface cover, along with additional channelization of rivers, will only serve to exacerbate the problem and increase runoff even further. Thus, the number of flooding events is expected to increase markedly in the future. While the number of extreme precipitation events is likely to increase, periods of light precipitation are expected to decrease, and in fact, this pattern has already been observed in some locations (Wu 2015). Further, there is evidence suggesting that seasonal precipitation patterns will also change, causing wet seasons to become wetter and dry seasons to become dryer (Chou and Lan 2012; Feng et al. 2013). It is important to note that globally, flooding is already the leading weather-related killer, and the number of people susceptible to flooding has increased dramatically in recent years as populations continue to migrate to coastlines and floodplains. Thus, the combined impact of climate change and population distribution has the potential to put many more people at risk of severe flooding in the future.

2.5 Drought

Severe drought already has the ability to have profound impacts on the security environment, with recent droughts playing a role in migration, famine, and possibly even war (Kebbede and Jacob 1988; Kelley et al. 2015). Most scientists believe climate change will act to enhance both the frequency and severity of droughts, particularly in marginal climate zones already susceptible to dry conditions (Feng and Fu 2013; Huang et al. 2017b). The factors responsible for this expected increase in drought conditions are complex and go beyond a simple higher temperature / higher evaporation relationship. First, the increase in extreme precipitation events and corresponding decrease in mild precipitation events have the ability to increase runoff and reduce moisture percolation into the soil. Further, the rise in temperature will likely reduce snowpack, a valuable resource many communities depend upon as a reliable source of water (Adam et al. 2009). Finally, a possible change in storm track and moisture transport also has the potential to reduce rainfall in some semi-arid regions (Hagos and Cook 2008).

While there is general agreement within the scientific community that climate change will likely amplify drought conditions particularly in arid and semi-arid regions, many scientists question the magnitude of this impact and believe the link between climate change and drought is often over–estimated. For example, while some argue climate change played a sizeable role in the 2011–2016 California drought (AghaKouchak et al. 2014), others dispute this claim and suggest any role of climate change was minimal (Mao et al. 2015). Similarly, the exact impact of climate change on the 2006–2010 drought across Syria has also been disputed. Some scientists suggest climate change played a major role in drought and migration across the region, but others conclude climate change had very little, if any impact (Kelley et al. 2015; Selby et al. 2017). To summarize, although many questions remain, it is clear that climate change has the potential to exacerbate the frequency and severity of drought, particularly across regions already prone to drought conditions.

2.6 Heat

Unlike most natural disasters which can lead a trail of destruction across a wide area, heat has little discernible impact on local infrastructure, and is often referred to as a "silent killer." Heat is usually considered the leading weather-related killer across the United States, and recent heat waves around the world have been responsible for tens of thousands of deaths, predominately in areas with highly variable climates. This variability in temperature is often as important as its intensity, if not more so, in determining a negative human health outcome since the human body is less able to adapt to extreme conditions. It is estimated that the 2010 heat wave across Russia was responsible for over 50,000 deaths, while the 2003 European heat wave killed over 40,000. (García-Herrera et al. 2010; Revich 2011). More recently, the 2015 Indian heat wave killed thousands more. Despite these staggering death tolls, the true impact of heat on human health is often underestimated; the increased stress on the human body can lead to increases in heart attack, stroke, and respiratory distress, sources of morbidity and mortality often not recorded as being directly heat-related.

There is strong scientific evidence that climate change will result in more severe, longer lasting, and more frequent heat waves in the future as a result of rising surface air temperatures (Meehl and Tebaldi 2004; Luber and McGeehin 2008). Further, there are indications that variability in temperatures will also increase in regions most susceptible to heat-related illness. A shift in both the intensity of heat, coupled with an increase in variability, has the potential to greatly increase the number of truly exceptional heat events.

Despite the more extreme conditions likely in the future, questions remain concerning how humans will respond. While some scientists suggest heat-related mortality will increase dramatically, others note that current heat-related illness is on the decline, particularly in locales with available air conditioning and other forms of mitigation (Bobb et al. 2014; Hajat et al. 2014; Gasparrini et al. 2015). Thus, while scientists are largely in agreement concerning the characteristics and frequency of future heat waves, the impact on human health remains unclear.

Literature Cited

Adam JC, Hamlet AF, Lettenmaier DP (2009) Implications of global climate change for snowmelt hydrology in the twenty-first century. Hydrol Process 23(7):962–972

AghaKouchak A, Cheng L, Mazdiyasni O, Farahmand A (2014) Global warming and changes in risk of concurrent climate extremes: insights from the 2014 California drought. Geophys Res Lett 41(24):8847–8852

Beckley B, Zelensky NP, Holmes SA, Lemoine FG, Ray RD, Mitchum GT, Desai S, Brown ST (2017) Global mean sea level trend from integrated multi-mission ocean altimeters TOPEX/Poseidon Jason-1 and OSTM/Jason-2 Version 4. Ver. 4. PO.DAAC, CA, USA. https://doi.org/10.5067/GMSLM-TJ124. Accessed 15 Oct 2017

Bender MA, Knutson TR, Tuleya RE, Sirutis JJ, Vecchi GA, Garner ST, Held IM (2010) Modeled impact of anthropogenic warming on the frequency of intense Atlantic hurricanes. Science 327(5964):454–458

Bengtsson L, Botzet M, Esch M (1996) Will greenhouse gas-induced warming over the next 50 years lead to higher frequency and greater intensity of hurricanes? Tellus A 48(1):57–73

Bengtsson L, Hodges KI, Roeckner E (2006) Storm tracks and climate change. J Clim 19(15):3518–3543

Bobb JF, Peng RD, Bell ML, Dominici F (2014) Heat-related mortality and adaptation to heat in the United States. Environ Health Perspect 122(8):811–816

Brooks HE (2013) Severe thunderstorms and climate change. Atmos Res 123:129–138

Chou C, Lan CW (2012) Changes in the annual range of precipitation under global warming. J Clim 25(1):222–235

Couvillion BR, Barras JA, Steyer GD, Sleavin W, Fischer M, Beck H, Trahan N, Griffin B, Heckman D (2011) Land area change in coastal Louisiana from 1932 to 2010. US Geological Survey Scientific Investigation Map 3164, scale 1:265,000

Dwyer JG, Camargo SJ, Sobel AH, Biasutti M, Emanuel KA, Vecchi GA, Zhao M, Tippett MK (2015) Projected twenty-first-century changes in the length of the tropical cyclone season. J Clim 28(15):6181–6192

Elsner JB, Elsner SC, Jagger TH (2015) The increasing efficiency of tornado days in the United States. Clim Dyn 45(3–4):651–659

Ezer T, Atkinson LP, Corlett WB, Blanco JL (2013) Gulf Stream's induced sea level rise and variability along the US mid-Atlantic coast. J Geophys Res Oceans 118(2):685–697

Feng S, Fu Q (2013) Expansion of global drylands under a warming climate. Atmospheric Chemistry Physics 13(19):14637–14665

Feng X, Porporato A, Rodriguez-Iturbe I (2013) Changes in rainfall seasonality in the tropics. Nat Clim Chang 3(9):811–815

García-Herrera R, Díaz J, Trigo RM, Luterbacher J, Fischer EM (2010) A review of the European summer heat wave of 2003. Crit Rev Environ Sci Technol 40(4):267–306

Gasparrini A, Guo Y, Hashizume M, Kinney PL, Petkova EP, Lavigne E, Zanobetti A, Schwartz JD, Tobias A, Leone M, Tong S (2015) Temporal variation in heat–mortality associations: a multicountry study. Environ Health Perspect 123(11):1200–1207

Groisman PY, Knight RW, Karl TR, Easterling DR, Sun B, Lawrimore JH (2004) Contemporary changes of the hydrological cycle over the contiguous United States: trends derived from in situ observations. J Hydrometeorol 5(1):64–85

Hagos SM, Cook KH (2008) Ocean warming and late-twentieth-century Sahel drought and recovery. J Clim 21(15):3797–3814

Haigh ID, Wahl T, Rohling EJ, Price RM, Pattiaratchi CB, Calafat FM, Dangendorf S (2014) Timescales for detecting a significant acceleration in sea level rise. Nat Commun 5:1–11

Hajat S, Vardoulakis S, Heaviside C, Eggen B (2014) Climate change effects on human health: projections of temperature-related mortality for the UK during the 2020s, 2050s and 2080s. J Epidemiol Community Health 68(7):641–648

Huang B, Thorne PW, Banzon VF, Boyer T, Chepurin G, Lawrimore JH, Menne MJ, Smith TM, Vose RS, Zhang H (2017a) NOAA Extended Reconstructed Sea Surface Temperature (ERSST), Version 5. NOAA National Centers for Environmental Information

Huang J, Yu H, Dai A, Wei Y, Kang L (2017b) Drylands face potential threat under 2 ° C global warming target. Nat Clim Chang 7:417–422

Intergovernmental Panel on Climate Change (IPCC) (2013) Climate change 2013: the physical science basis. Contribution of working group I to the fifth assessment report of the Intergovernmental Panel on Climate Change. Cambridge University Press, Cambridge

Joughin I, Smith BE, Medley B (2014) Marine ice sheet collapse potentially under way for the Thwaites Glacier Basin, West Antarctica. Science 344(6185):735–738

Kang NY, Elsner JB (2015) Trade-off between intensity and frequency of global tropical cyclones. Nat Clim Chang 5:661–664

Kebbede G, Jacob MJ (1988) Drought, famine and the political economy of environmental degradation in Ethiopia. Geography 73(1):65–70

Kelley CP, Mohtadi S, Cane MA, Seager R, Kushnir Y (2015) Climate change in the Fertile Crescent and implications of the recent Syrian drought. Proc Natl Acad Sci 112(11):3241–3246

Knutson TR, McBride JL, Chan J, Emanuel K, Holland G, Landsea C, Held I, Kossin JP, Srivastava AK, Sugi M (2010) Tropical cyclones and climate change. Nat Geosci 3(3):157–163

Kossin JP, Emanuel KA, Vecchi GA (2014) The poleward migration of the location of tropical cyclone maximum intensity. Nature 509:349–352
Luber G, McGeehin M (2008) Climate change and extreme heat events. Am J Prev Med 35(5):429–435
Mann ME, Emanuel KA (2006) Atlantic hurricane trends linked to climate change. EOS Trans Am Geophys Union 87(24):233–241
Mao Y, Nijssen B, Lettenmaier DP (2015) Is climate change implicated in the 2013–2014 California drought? A hydrologic perspective. Geophys Res Lett 42(8):2805–2813
Mears CA, Wentz FJ (2017) A satellite-derived lower-tropospheric atmospheric temperature dataset using an optimized adjustment for diurnal effects. J Clim 30(19):7695–7718
Meehl GA, Tebaldi C (2004) More intense, more frequent, and longer lasting heat waves in the 21st century. Science 305(5686):994–997
National Aeronautics and Space Administration (NASA) (2017) Global land-ocean temperature index. http://climate.nasa.gov/system/internal_resources/details/original/647_Global_Temperature_Data_File.txt. Accessed 15 Oct 2017
National Oceanic and Atmospheric Administration (NOAA) (2017a) National centers for environmental information, climate at a glance: global time series. http://www.ncdc.noaa.gov/cag/. Accessed 15 Oct 2017
National Oceanic and Atmospheric Administration (NOAA) (2017b) National centers for environmental information. U.S. Tornado climatology historical records and trends. https://www.ncdc.noaa.gov/climate-information/extreme-events/us-tornado-climatology/trends. Accessed 15 Oct 2017
National Snow and Ice Data Center (NSIDC) (2017) Sea ice index, Version 2. http://nsidc.org/data/docs/noaa/g02135_seaice_index/index.html. Accessed 15 Oct 2017
Nerem RS, Chambers D, Choe C, Mitchum GT (2010) Estimating mean sea level change from the TOPEX and Jason Altimeter Missions. Mar Geod 33(1):435 2016
Oouchi K, Yoshimura J, Yoshimura H, Mizuta R, Kusunoki S, Noda A (2006) Tropical cyclone climatology in a global-warming climate as simulated in a 20 km-mesh global atmospheric model: frequency and wind intensity analyses. J Meteorol Soc Jpn 84(2):259–276
Revich BA (2011) Heat-wave, air quality and mortality in European Russia in summer 2010: preliminary assessment. Ekologiya Cheloveka/Human Ecology 7:3–9
Rignot E, Velicogna I, van den Broeke MR, Monaghan A, Lenaerts JT (2011) Acceleration of the contribution of the Greenland and Antarctic ice sheets to sea level rise. Geophys Res Lett 38(5)
Root TL, Price JT, Hall KR, Schneider SH, Rosenzweig C, Pounds JA (2003) Fingerprints of global warming on wild animals and plants. Nature 421(6918):57–60
Sallenger AH, Doran KS, Howd PA (2012) Hotspot of accelerated sea-level rise on the Atlantic coast of North America. Nat Clim Chang 2(12):884–888
Santer BD, Mears C, Wentz FJ, Taylor KE, Gleckler PJ, Wigley TML, Barnett TP, Boyle JS, Brüggemann W, Gillett NP, Klein SA (2007) Identification of human-induced changes in atmospheric moisture content. Proc Natl Acad Sci 104(39):15248–15253
Saunders MA, Lea AS (2008) Large contribution of sea surface warming to recent increase in Atlantic hurricane activity. Nature 451(7178):557–560
Scambos TA, Bell RE, Alley RB, Anandakrishnan S, Bromwich DH, Brunt K, Christianson K, Creyts T, Das SB, DeConto R, Dutrieux P (2017) How much, how fast?: a science review and outlook for research on the instability of Antarctica's Thwaites Glacier in the 21st century. Glob Planet Chang 153:16–34
Selby J, Dahi OS, Fröhlich C, Hulme M (2017) Climate change and the Syrian civil war revisited. Polit Geogr 60:232–244
Sella GF, Stein S, Dixon TH, Craymer M, James TS, Mazzotti S, Dokka RK (2007) Observation of glacial isostatic adjustment in "stable" North America with GPS. Geophys Res Lett 34(2)
Shiu CJ, Liu SC, Fu C, Dai A, Sun Y (2012) How much do precipitation extremes change in a warming climate? Geophys Res Lett 39(17)

Spencer RW, Christy JR, and Braswell WD (2015) Version 6.0 of the UAH temperature dataset released: new LT trend = +0.11 C/decade. http://www.drroyspencer.com/2015/04/version-6-0-of-the-uah-temperature-dataset-released-new-lt-trend-0-11-cdecade/. Accessed 15 Oct 2017

Trapp RJ, Diffenbaugh NS, Brooks HE, Baldwin ME, Robinson ED, Pal JS (2007) Changes in severe thunderstorm environment frequency during the 21st century caused by anthropogenically enhanced global radiative forcing. Proc Natl Acad Sci 104(50):19719–19723

Trenberth KE (2011) Changes in precipitation with climate change. Clim Res 47(1/2):123–138

Walsh KJ, McBride JL, Klotzbach PJ, Balachandran S, Camargo SJ, Holland G, Knutson TR Kossin JP, Lee TC, Sobel A, Sugi M (2016) Tropical cyclones and climate change. Wiley Interdiscip Rev Clim Chang 7(1):65–89

Walther GR, Post E, Convey P, Menzel A, Parmesan C, Beebee TJ, Fromentin JM, Hoegh-Guldberg O, Bairlein F (2002) Ecological responses to recent climate change. Nature 416(6879):389–395

Webster PJ, Holland GJ, Curry JA, Chang HR (2005) Changes in tropical cyclone number, duration, and intensity in a warming environment. Science 309(5742):1844–1846

Wu SY (2015) Changing characteristics of precipitation for the contiguous United States. Clim Chang 132(4):677–692

Xu J, Grumbine RE, Shrestha A, Eriksson M, Yang X, Wang YUN, Wilkes A (2009) The melting Himalayas: cascading effects of climate change on water, biodiversity, and livelihoods. Conserv Biol 23(3):520–530

Abrupt Climate Change

Francis A. Galgano

Abstract Changes in Holocene climate have generally been gradual and the environmental security effects have been manageable. However, the long–term Holocene climate record also indicates that significant change is usually abrupt with major implications for regional and global stability. Economic and ecological impacts of abrupt climate events could exceed the adaptation capacity of most states. Climate models suggest that a potential abrupt event may result in harsher winter conditions, severe drought, and more intense zonal weather patterns. These conditions can induce major food and water shortages, persistent epidemics, and disputed access to energy resources. This scenario has the clear potential to destabilize the geopolitical environment thus leading to violent interstate conflict. This chapter will examine an abrupt climate change scenario from an environmental security perspective and present a regional framework to demonstrate the spatial pattern of potential threats.

Keywords Abrupt climate change · Adaptation · Akkadian · Anasazi · Climate change · Climate models · Conflict · Darfur · Developing states · Drought · Dynamic equilibrium · Environmental degradation · Environmental security · Fertile Crescent · Gulf stream · Holocene · Ice sheet Core · Little ice age · Mega drought · Mesopotamia · Rwanda · Static equilibrium · Steady state equilibrium · Syria · Thermohaline conveyor · Younger Dryas

1 Introduction

Climate data suggest the contemporary late Holocene warming trend will continue well into the twenty-first Century. Nevertheless, in the context of the entire Holocene record, changes in global climate since 1850 have been gradual and rather unremarkable in speed and magnitude; and the environmental security effects of global warming are expected to be manageable for most states (Fig. 1) (NRC 2002). More importantly, there is a growing body of evidence indicating that the current warming

F. A. Galgano (✉)
Department of Geography and the Environment, Villanova University, Villanova, PA, USA
e-mail: francis.galgano@villanova.edu

F. Galgano (ed.), *The Environment-Conflict Nexus*, Advances in Military Geosciences, https://doi.org/10.1007/978-3-319-90975-2_4

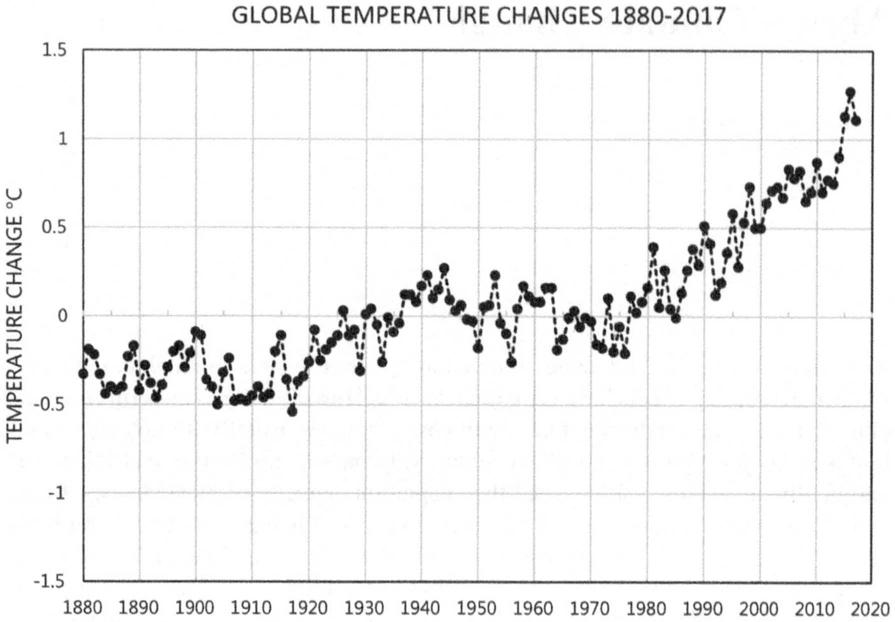

Fig. 1 The contemporary warming trend following the Little Ice Age (i.e. "global warming"). The rate of change since 1850 has been + 0.23 °C per decade

trend could trigger a slowing of the ocean's Thermohaline Conveyor, potentially initiating a rapid downward adjustment in global temperature patterns. This change could be catastrophic and lead to severe winters and major drought in the world's principal food producing regions (Alley and Pugh 2001; Gagosian 2003; USGS 2008a). This potential abrupt climate shift is problematical because it can be manifested as a rapid and pervasive global cooling and drying trend that persists for many decades, and transpire at magnitudes that defy human adaptation capacity (Schwartz and Randall 2003; USGS 2008b).

It is conceivable that anthropogenic influences on global climate may increase the likelihood of an abrupt cooling event (Joyce 2003). This is important because the economic and ecological impacts of such events could be catastrophic. A number of models have been developed that indicate that an abrupt change in climate can result in rapid and substantial cooling; perhaps by as much as 9–18 °C in a single decade (USGS 2009). These conditions may induce acute food and water shortages, epidemics, mass migrations, and disputed access to energy supplies, thus potentially leading to violent interstate conflict. Climate change is a persistent reality and is significant because many conflicts are triggered by the exposure of a society to environmental effects. Historical records indicate that abrupt climate changes have occurred during the recent past at regional scales and may be linked to violent conflict (Diamond 2005). Hence, an understanding of the environment–conflict nexus offers an especially valuable vantage point from which to conduct an analysis of conflict related to environmental security. This chapter examines an abrupt climate change scenario patterned after the so–called 8.2 k climate event, which took place during the recent geologic past, and presents a regional framework to demonstrate the spatial pattern of changes in climate and potential threats.

Environmental stress will play a pervasive role in future conflicts because the economic well–being of about half of the world's population is tied directly to the land. This is important because almost 75% of the world's most impoverished inhabitants are subsistence farmers (Homer-Dixon 1999). Drought, desertification, deforestation, and soil exhaustion are major problems in these regions. This is a security concern because anticipated population growth, especially in the developing world, will result in higher *per capita* consumption rates. Consequently, we should anticipate that environmental stress would make an increasingly significant contribution to three modes of conflict: (1) ethnic/racial warfare enabled by environmental stress and population pressure; (2) civil warfare instigated by economic collapse; and (3) limited-scale interstate wars.

Environmental security is a process involving, environmental risk analysis based on natural processes that destabilize the environment and contribute to instability or conflict. The fundamental components of environmental security include environmental processes that undermine governments, promote instability, and trigger violent conflict. Since 1990, violent conflicts have occurred in many places and it would be too simplistic, and probably incorrect, to assert that environmental stress instigated each of these conflicts, and too difficult to disaggregate their human and environmental components (Homer-Dixon 1999).

Nevertheless, relationships between food, population, climate, resources, and conflict are evident in many developing states (Femina and Werrell 2012). This apparent Malthusian paradigm generates a great deal of disagreement among scholars. However, all factions have to agree on an undeniable outcome that has been evident in places like Darfur, Rwanda, and now Syria: that is, population growth and environmental stress, superimposed over latent ethic and political divisions will in the end, be solved one-way or another. Detractors of the environment–conflict nexus argue that wars result solely from politico-military factors, and are minimally influenced by environmental stress at best, and perhaps allude to environmental determinism. However, environmental stress results from the combined influence of anthropogenic effects on the environment in conjunction with the vulnerability of the ecosystem. Scarcity and stress contribute to conflict only under certain circumstances, but there is no deterministic link. Clearly, not all violent episodes are alike and the influence of environmental stress on conflict will vary in magnitude from example to example.

2 Historical Context

Environmental security is not a new concept, and history suggests that ancient cultures experienced violent conflict and collapse triggered by abrupt climate events. One such example is the Anasazi culture of the American Southwest, which flourished between 900 and 1100 AD, and then suddenly collapsed, coincident with a rapid drying trend (Mays 2007). Another compelling example is the demise of the ancient Akkadian Empire 4000 years ago. This civilization developed in Mesopotamia in what is now Iraq (Gibbons 1993). The Akkadian Empire prospered as an agrarian society between 2300 and 2200 BC, and archeological evidence suggests that it came to a sudden halt in 2200 BC. Weiss and Bradley (2001) suggest that Akkadian society literally dried up from an abrupt climate event. This shift in precipitation patterns caused a severe drought, which advanced from north to south along the Fertile Crescent. As a result, tens of thousands of people from the northern region migrated to southern cities, overtaxing precarious water and food supplies. Archeological evidence suggests that chaos ensued and the southerners made war on their unwanted northern cousins. Weiss indicates, "*... for the first time, we've indentified abrupt climate change directly linked to the collapse of a thriving civilization*" (Weiss and Bradley 2001, p. 611). Both examples illustrate the potential effects of the environment–conflict nexus. They demonstrate how environmental degradation triggered by an abrupt climate event can erode a society's ability to assure the most essential aspects of security because environmental change undermines natural support systems on which humans depend. Exposure to environmental degradation has the potential to erode the resource base and make it impossible for a regime to adapt and provide for the population.

3 Abrupt Climate Change

Quaternary climatologists have long been familiar with the variable and sometimes volatile nature of Earth's climate record. However, the dominant popular perspective of climate change is that it changes gradually and that somehow the recent warming trend is fundamentally anomalous, notwithstanding significant geologic evidence to the contrary. Geologic and fossil records clearly suggest that climate actually undergoes large magnitude changes sometimes with astonishing speed. Although the Holocene climate record has been the most stable in recent geologic history, abrupt changes in climate appear to be the norm (Alley and Pugh 2001).

Abrupt climate changes are oscillations in climate, typically by 9–18 °C, that occur over geologically short time spans, and three such Holocene events are illustrated in Fig. 2 (NCDC 2008c). Several definitions of abrupt climate change have been offered in the literature. The National Research Council (NRC 2002) defines it as fluctuations in climate that take place when the global climate system is forced across a threshold, thus causing a transition to a new climate state at a rate faster than the initial cause. The U.S. Climate Change Science Program (USGS 2009) defines it as a change that occurs over a few decades, endures for several decades, and causes substantial potential disruptions in human and natural systems.

Rapid changes in climate take place because the climate system is non–linear. That is inputs and outputs are not proportional and once the system crosses a threshold, change is characteristically abrupt, rather than slow and gradual; hence, the

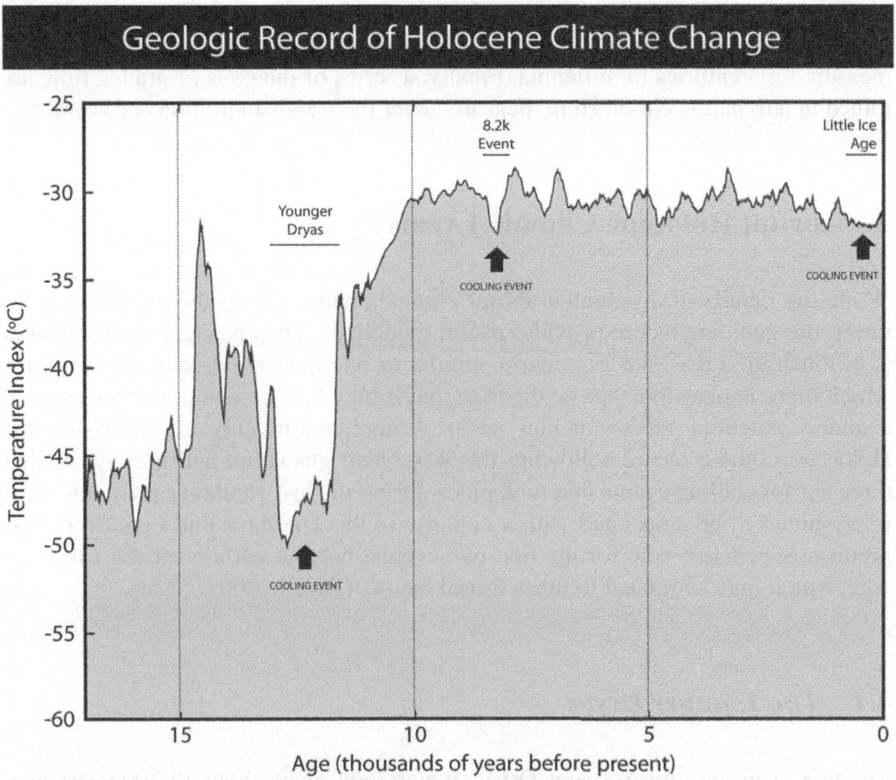

Fig. 2 The Holocene warming trend manifests the variability of climate conditions and three so-called abrupt events since 12,000 years B.P. (USGS 2009)

system exhibits multiple equilibria (Schwartz and Randall 2003). Hence, the concept of equilibrium is a seminal issue in understanding climate change. The inherent problem is that climate is assumed to be at some equilibrium and thus, the recent warming trend is fundamentally abnormal implying that climate has persisted in an unchanging condition. In fact, variability is the norm as suggested by the Holocene climate record (Fig. 2) and the meaning of equilibrium is dependent on the time scales being considered. There are different types of equilibrium, and indeed, different forms of equilibrium are related to specific intervals of time. An abrupt change occurs when climate is pushed beyond a threshold, precipitating change from one equilibrium state to another (Gagosian 2003).

Given that multiple equilibria exist, the climate record is necessarily highly variable with cyclical events occurring at different scales and magnitudes. The concept of static equilibrium explains conditions in the climate system that persists over short, or steady, time intervals of years and decades. An example of static equilibrium is the periodic warm and cool periods that persist for a decade, such as the warm period of the 1930s and cool period of the 1940s (Fig. 1). A number of static equilibria are linked to compose steady state equilibrium—also known as graded time—with periodic fluctuations above and below an average condition typically measured in centuries to millennia. Finally, a series of intervals of graded time are joined in a dynamic equilibrium measured over thousands to millions of years.

4 Abrupt Holocene Climate Events

While the details of a potential abrupt climate change cannot be predicted accurately, the geologic record provides useful guidelines. The objective of this chapter is to illustrate a reasonable scenario, similar to one that has already occurred, for which there is plausible data to theorize that it may happen again, so that we may examine potential environmental security implications. Fig. 2 illustrates the Holocene climate record following the Wisconsin glaciation and it suggests that three abrupt cooling events that took place during the last 15,000 years. Each event is presumed to be associated with a collapse in the Thermohaline Conveyor. This premise is certainly true for the first two events, but less certain for the Little Ice Age, which may be related to other causal factors (Fagan 2000).

4.1 The Younger Dryas

The first event was the Younger Dryas, which took place about 12,700 years ago. Paleoclimate data indicates that this cooling event, which lasted 1300 years, was characterized by a rapid drop in temperatures in Greenland, and substantial changes throughout the North Atlantic region (NCDC 2008b). The striking feature of the Younger Dryas was that it resulted in a series of decadal drops in temperature of around 9 °C each, and the extreme cold and dry conditions persisted for a millennia. While this event had an enormous effect throughout the North Atlantic region, its impact would be more severe today in our densely populated and urbanized society. In fact, it would be fair to say that a similar occurrence, today, would be a civilization changing event (Pearce 2007). However, an event of this magnitude is less likely because similar physical conditions (i.e., massive volumes of glacial melt water) are not present today.

4.2 The 8.2 K Climate Event

The climate change scenario outlined in this paper is modeled after the so-called 8.2 k climate event, which was a century–long episode (Fig. 2). The Paleoclimate record suggests that this scenario is perhaps the most plausible. This event followed an extended period of warming—much like the period we appear to be in today—followed by a sudden, drastic cooling (NCDC 2008a). In this scenario, Greenland's average annual temperatures dropped by roughly 12 °C, and similar temperature decreases are likely to have occurred throughout the North Atlantic region. During this climate event severe winters in Europe and other regions adjacent to the North Atlantic caused glaciers to advance, rivers to freeze, and agricultural lands to be far less productive. Paleoclimatic evidence suggests strongly that this event was associated with a collapse of the Thermohaline Conveyor (NCDC 2008a).

4.3 The Little Ice Age

Beginning in the fourteenth century, the North Atlantic region experienced a dramatic cooling period that lasted until about 1850, which is commonly known as the Little Ice Age. This event may have been caused by a slowing of the ocean conveyor, although it is more generally thought that reduced solar output and/or volcanic eruptions may have contributed to this climate event (Fagan 2000). The Little Ice Age was a period of anomalous cooling that occurred following the Medieval Warm Period, but is not a true ice age in a literal sense. There is a general lack of agreement among climatologists and historians on the precise start and end dates of this event, nevertheless is conventionally defined as a period extending from about 1300 to about 1850. The Little Ice Age is important in the context of this study because it caused severe winters and profound agricultural, economic, and political impacts to Europe. It significantly destabilized much of Europe by causing famines, epidemics and political instability. Finally, it was a period of persistent warfare. While climate crises like the Little Ice Age are not solely responsible for violent conflict, it is undeniable that they have a large impact on society (Hulme 2003).

5 The Thermohaline Conveyor

One important question then is what mechanism could possibly drive such large–scale abrupt changes in climate? So far, scientists have recognized only one viable process–response mechanism: a rapid restructuring of global ocean currents. These currents, collectively known as the Thermohaline Conveyor, distribute immense volumes of heat around the planet, and therefore play an essential role in regulating the climate system (Calvin 1998). A simplified explanation of the basic ocean-atmosphere dynamics that are regulated by the Thermohaline Conveyor is useful to explain this process-response mechanism (Fig. 3).

Equatorial ocean waters are warmed by solar radiation, which creates a surplus of heat, and enhances evaporation in the tropics; thus leaving tropical ocean water saltier. The Gulf Stream, an extension of the Thermohaline Conveyor, carries an enormous volume of warm, salty water to the North Atlantic region. This oceanic

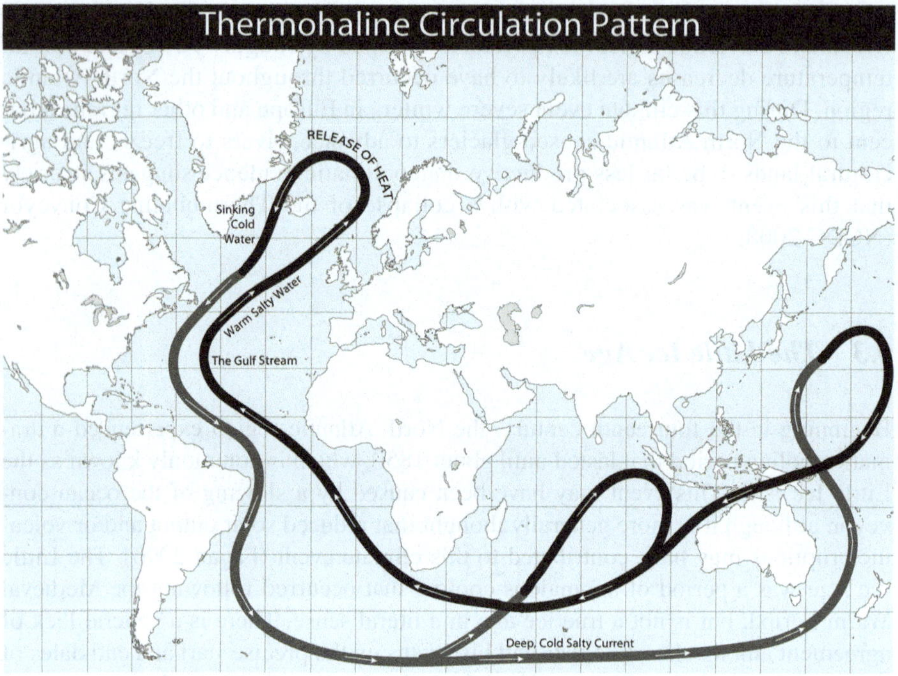

Fig. 3 The ocean's Thermohaline Conveyor delivers warm, salty water to the North Atlantic region. Heat is released, thus moderating Europe's climate, especially in winter (USGS 2009)

heat pump is an important mechanism for balancing the global energy budget and moderating climate in the North Atlantic region. At northern latitudes, the ocean releases heat to the atmosphere—especially in winter—hence, it warms North Atlantic regions by as much as 5–7 °C and significantly moderates average winter temperatures (USGS 2009).

However, records of past climates from sources such as deep–sea sediments and ice–sheet cores, suggest that the Thermohaline Conveyor has shut down several times in the recent geologic past (Alley and Pugh 2001). This shutdown limited heat delivery to the North Atlantic, triggering a substantial cooling trends throughout the region. The crucial component of the Thermohaline Conveyor is water saltiness. For an assortment of reasons, North Atlantic waters are relatively salty. This system functions because salty water is denser than fresh water; and cold water is denser than warm. Consequently, when the warm, salty waters of the North Atlantic release heat to the atmosphere, they become colder and sink. Once this happens, a massive volume of cold water sinks to the abyss. This water flows at great depths throughout the world's oceans (Calvin 1998).

However, this process can be dislocated if the North Atlantic's cold, salty waters do not sink. If that happens, the principal process driving global ocean circulation could taper off and potentially cease altogether (NRC 2002). The resulting restructuring of the ocean's currents could significantly modify global climate patterns.

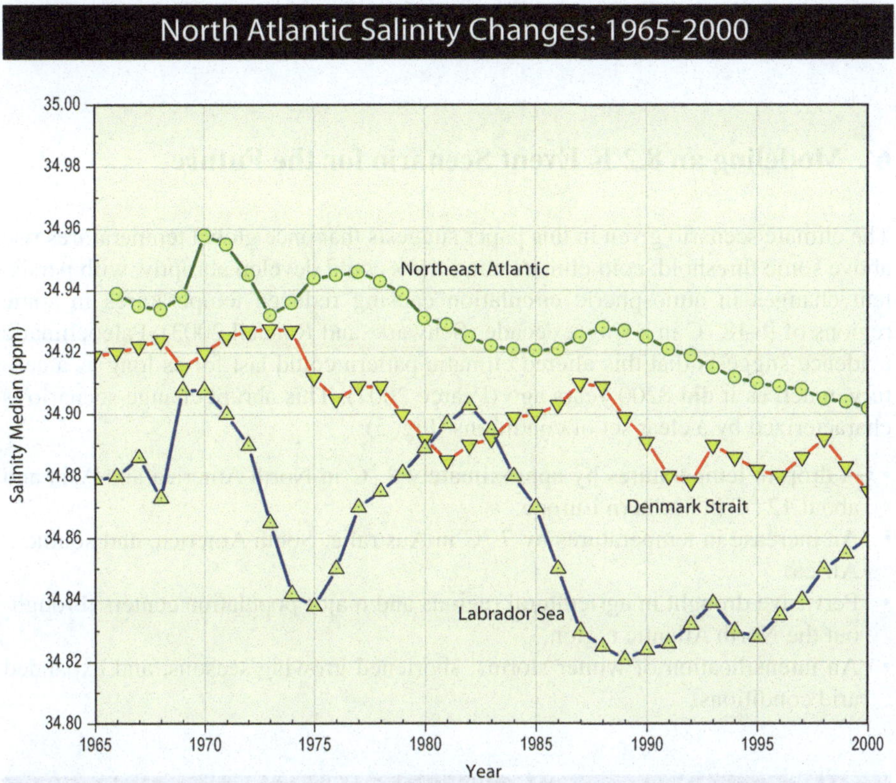

Fig. 4 Data indicating that the surface waters of the North Atlantic are becoming fresher and perhaps establishing conditions for a disruption of the Thermohaline conveyor (after Dickson et al. 2002)

This can happen if an influx of fresh water into the North Atlantic could create a cap of buoyant fresh water over the denser, saltier water. This fresh water would in effect insulate the surface, inhibiting the transfer of heat to the atmosphere. At the same time, the influx of fresh water could dilute the North Atlantic's salinity. Thus, at a critical but unknown threshold, when North Atlantic waters are no longer sufficiently salty and dense, they may stop sinking, thus shutting down the Thermohaline Conveyor (USGS 2009).

Dickson et al. (2002) presented data that tend to support this model. These oceanographers, monitoring conditions in the North Atlantic, concluded that it has been steadily freshening for the past 40 years but especially during the past decade. These data (Fig. 4) suggests that since the mid–1960s, North Atlantic waters have steadily and noticeably become less salty. Therefore, the crucial question is at what threshold will the conveyor cease to operate? The short answer is we just do not know, and furthermore, scientists have not convincingly determined the relative contributions of sources that may be adding fresh water to the North Atlantic.

Among suspected sources are melting glaciers, increased precipitation, or rivers that discharge into the Arctic Ocean (Dickson et al. 2002).

6 Modeling an 8.2 K Event Scenario for the Future

The climate scenario given in this paper suggests that once global temperatures rise above some threshold, cold climate conditions could develop abruptly, with persistent changes in atmospheric circulation causing reduced temperatures in some regions of 9–18 °C in a single decade (Schwartz and Randall 2003). Paleoclimatic evidence suggests that this altered climatic pattern could last for as long as a century, much as it did 8200 years ago (Pearce 2007). This abrupt change scenario is characterized by a clear set of conditions (Fig. 5):

- A drop in temperatures by approximately 9 °C in North America and Asia and about 12 °C in northern Europe.
- An increase in temperatures by 7 °C in Australia, South America, and southern Africa.
- Pervasive drought in agricultural regions and major population centers throughout the North Atlantic region.
- An intensification of winter storms, shortened growing seasons, and expanded arid conditions.

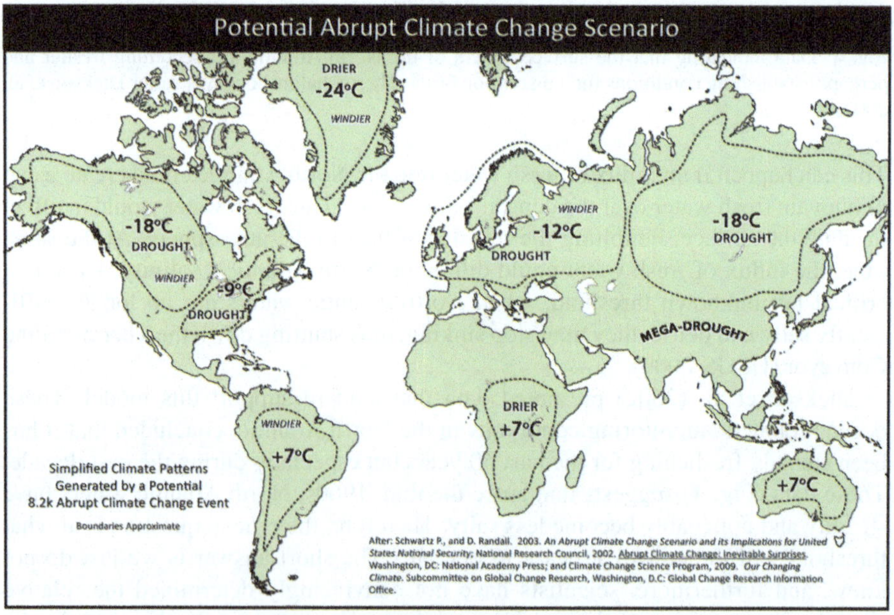

Fig. 5 An abrupt climate change scenario modeled after a potential 8.2 k event

This scenario suggests that such an abrupt climate change could potentially destabilize the geopolitical environment, leading to violent conflict resulting from resource constraints such as severe food and water shortages. These climate–related events would also have a substantial effect on society. For example, the 1998 U.S. drought cost it $40 billion in lost revenue. The drought that persisted in the Horn of Africa from 1984–1985 is estimated to have led to the death of 750,000 people (USGS 2008b; Galgano 2013).

However, the 8.2 k scenario presented in this paper will certainly be much longer with larger scale implications than these two examples. For instance, the 8.2 k climate conditions may reduce rainfall leading to a possible 15% decline in grass productivity, causing reductions in cattle weight by as much as 12%, thus reducing protein supplies. Furthermore, milk production is expected to decrease by some 25–30%, and new pests are likely to spread in agricultural regions, further reducing food production. These conditions could beset the world's primary food–producing regions at once, thus challenging the idea that society and governments can adapt (Schwartz and Randall 2003).

This 8.2 k climate scenario may result in significantly reduced carrying capacity, and mounting tensions could lead to environmentally triggered warfare. Given that more than 400 million people live in drier, over–populated and economically poor regions; this scenario and its resultant effects pose a severe risk to political, economic, and social stability especially in South America, Africa, the Middle East, and South Asia (Pearce 2007). In the developing world, where poorly governed states lack the resources required to adapt to such severe conditions, the problem is likely to be exacerbated. States with adequate resources may resort to enforced isolation to preserve resources. Less fortunate states may instigate warfare to secure access to food, water, or energy. Thus, national defense priorities could shift with a new goal of securing resources for survival.

Abrupt climate change events that have occurred in the past suggest that it is prudent to envisage such a scenario for the future as reasonable, especially because data indicate that we could be on the cusp of such an event. This scenario makes reasonable assumptions about which regions are likely to be colder, drier, and windier. Nevertheless, there is no way to confirm its assumptions because the most complex climate models cannot estimate in detail how climate change will unfold, how discrete regions will be affected, and how governments might respond.

This scenario estimates that average annual rainfall will decline by nearly 30% and winds will become about 15% stronger. These persistently colder and arid conditions are assumed to be more severe in continental interiors. The effects on populations would be devastating as persistent drought in the world's most populated regions lingers for decades. Given that we expect to experience a considerable reduction in precipitation, water supplies will quickly be over-taxed and reserves will be depleted to the extent that shortages will overwhelm conservation options (Joyce 2003).

The fundamental problem presented by this scenario is that cooling will be more pronounced during the winter season and will persist throughout the year. Furthermore, winters are expected to become increasingly intense and less predictable, thus making growing seasons shorter. Added to the problem of intensive cooling and summer dryness, it is projected that wind patterns and velocities will become more zonal: i.e., winds flow more frequently from the pole to the equator. This altered pattern implies that wind speeds will accelerate as the atmosphere tries to adjust dynamically to a stronger pole–to–equator temperature gradient. Consequently, cold air blowing across Europe, North America, and Asia could potentially generate harsh conditions for agriculture, and the combination of wind and dryness will cause widespread dust storms and soil loss. Given these conditions and the data in the geologic record, we can expect that by the end of the first decade of this event, the North Atlantic region's climate will approximate Siberia's (Crowley and North 1988).

7 Summary and Conclusions

The abrupt climate change scenario described in this paper has the potential to profoundly affect agriculture, biomes, water supplies, and energy in most of the world's major population centers. Crop yields, affected by frigid temperature and water stress, as well as truncated growing season are expected fall by 10–25%. They will be less predictable as key regions shift to a devastating cooling trend. This scenario also means that affected regions will experience more frequent epidemics from vectors and blowing dust, and arid conditions will introduce new varieties of agricultural pests. Developing states and states with limited resources may not have the means to adapt to these new conditions. One concern is that this scenario will induce mass migrations, such as those suggested in the Akkadian example leading to acute civil conflict and regional warfare.

The spread of human civilization flourished with the stabilization and warming of Earth's climate following the last glacial period. A warmer, stable climate meant that humans could develop agriculture and permanent settlements. With the end of the Younger Dryas and the warming and stabilization that followed, human civilization benefited from an extended period of relatively stable climate conditions. Modern civilization has never experienced weather conditions as persistently disruptive as the ones outlined in the 8.2 k scenario. Thus, the implications for national security outlined in this paper are only hypothetical. The tangible effects would of course vary greatly depending on the subtleties and abruptness of climate conditions, the ability of societies to adapt, and decisions by policymakers. Nonetheless, violence and disruption triggered by abrupt climate change pose a different type of threat to national security than we are accustomed to today. Military confrontation may be triggered by a need for natural resources and the shifting motivation for confrontation would alter which countries are most vulnerable.

Literature Cited

Alley RB, Pugh E (2001) The past says abrupt climate change is in our future. Science Daily, 14 December 2001. http://www.sciencedaily.com/releases/2001/011214081853.htm. Accessed 23 Feb 2012

Calvin WH (1998) The great climate flip-flop. The Atlantic online http://www.theatlantic.com/issues/98jan/climate.htm. Accessed 23 Feb 2015

Crowley TJ, North GR (1988) Abrupt climate change and extinction events in early history. Science 240(4855):996–1002

Diamond J (2005) Collapse. Penguin Books, New York

Dickson B, Yashayaev I, Meincke J, Turell B, Dye S, Holfort J (2002) Rapid freshening of the deep North Atlantic Ocean over the past four decades. Nature 416:832–837

Fagan B (2000) The little ice age. Basic Books, New York

Femina F, Werrell CE (2012) Update: climate and security 101: Why the U.S. national security establishment takes climate change seriously. The Center for Climate and Security, Briefer no. 23, 26 February 2014. https://climateandsecurity.org/reports/#BRIEFERS. Accessed 12 Jan 2017

Gagosian RB (2003) Abrupt climate change: should we be worried? Panel on abrupt climate change, World Economic Forum, Davos, Switzerland

Galgano FA (2013) Disasters and conflict: the Ogaden War of 1977. Pennsylvania Geogr 51(2):66–90

Gibbons A (1993) How the Akkadian empire was hung out to dry. Science, News Services, 261(5124):985

Homer-Dixon TF (1999) Environmental scarcity and violence. Princeton University Press, Princeton

Hulme M (2003) Abrupt climate change: can society cope? Phil Trans R Soc London 361(1810):2001–2021

Joyce T (2003) Are we on the brink of a new Little Ice Age? Woods Hole Oceanographic Institution. http://www.whoi.edu/page.do?pid. Accessed 21 Jan 2009

National Climate Data Center (NCDC) (2008a) Post-glacial cooling 8,200 years ago. NOAA Satellite and Information Service. http://www.ncdc.noaa.gov/paleo/abrupt/data5.html. Accessed: 23 Feb 2009

National Climate Data Center (NCDC) (2008b) Defining abrupt climate change. NOAA Satellite and Information Service. http://www.ncdc.noaa.gov/paleo/abrupt/story1.html. Accessed: 23 Feb 2012

National Climate Data Center (NCDC) (2008c) The Younger Dryas. NOAA Satellite and Information Service. http://www.ncdc.noaa.gov/paleo/abrupt/data4.html. Accessed: 23 Feb 2009

National Research Council (NRC) (2002) Abrupt climate change: inevitable surprises. National Academy Press, Washington, DC

Pearce F (2007) With speed and violence: why scientists fear tipping points in climate change. Beacon Press, Boston

Schwartz P, Randall D (2003) An abrupt climate change scenario and its implications for United States national security. Global Business Network. http://www.bbn.com/ArticleDisplayServlet.srv?aid=26231. Accessed 28 Aug 2015

United States Geological Survey (USGS) (2009) Abrupt climate change: Final report, synthesis and assessment product 3.4. U.S. Climate Change Science Program and the Subcommittee on Global Change Research, U.S. Government Printing Office, Washington, DC

U.S. Geological Survey (USGS) (2008a) Abrupt climate change. U.S. Climate Change Science Program and Subcommittee on Global Change Research. Global Change Research Information Office, Washington, DC. https://www.globalchange.gov/browse/reports/sap-34-abrupt-climate-change. Accessed 12 Jan 2013

U.S. Geological Survey (USGS) (2008b) Abrupt climate change: will it happen this century? Science Daily, 27 December 2008. http://www.sciencedaily.com/releases/2008/12/08122722550.htm. Accessed 23 Feb 2009

Weiss H, Bradley H (2001) What drives societal collapse? Science 291(5504):609–610

Water in the Middle East

Francis A. Galgano

Abstract Water resources are a particularly problematic variable in the environmental security milieu because water is an essential resource for which there is no substitute. The volume of renewable fresh water is finite and not equitably distributed in a spatial sense. From a geopolitical perspective, the world's largest river systems are shared by multiple states and the potential for conflict is high. Historically water conflicts have resolved by cooperative means and states have relied on technology, trade, and diplomatic solutions. This chapter argues that the security landscape has changed profoundly, and the history of cooperative water–conflict resolution is no longer a reliable guide to the future. Rather, the continued peaceful resolution of interstate water conflicts is inconsistent with the realities of the emerging national security landscape. Climate change is already adversely affecting the distribution of water in many critical water basins, and the simultaneous proliferation of failing states has reduced the potential for diplomatic resolutions. This chapter examines linkages between environmental stress, regional instability, water availability, and conflict and uses the Middle East as a case study to highlight these points.

Keywords Adaptation · Climate change · Cold War · Conflict · Desalinization · Environmental security · Environmental stress · Equitable apportionment · Euphrates river · Globalization · Governance · Irrigation · Jordan river · Middle East · National security · Nile river · Population · Population · Regional stability · Scarcity · Sovereignty · Tigris river · Transboundary water basin · Virtual water · Water · Water scarce · Water stressed

---

F. A. Galgano (✉)
Department of Geography and the Environment, Villanova University, Villanova, PA, USA
e-mail: francis.galgano@villanova.edu

F. Galgano (ed.), *The Environment-Conflict Nexus*, Advances in Military Geosciences, https://doi.org/10.1007/978-3-319-90975-2_5

1 Introduction

The nexus of the environment and conflict—that is, environmental security—has become an important paradigm in national security planning. Indeed, Kaplan (2000) suggests that water and other environmental factors represent the core foreign policy challenge in this century. The U.S. Department of Defense (DoD) reinforced this notion when it identified environmental factors and resource scarcity as important features of the national security landscape (DoD 2014). The DoD suggests that national security affairs may no longer just be about armies and weapons; but instead, climate, resources, and demographics may now be viewed as being equally important as traditional elements of national power (Butts 2011).

Environmental security refers to a broad range of security issues exacerbated by environmental factors and suggests that environmental stress has the potential to destabilize states and trigger violent conflict (Galgano and Krakowka 2011). Water is a particularly challenging factor in the environmental security milieu because it is an essential resource, and the fresh water supply problem only promises to intensify in a greenhouse world. Increases in global population and the attendant economic demands this growth engenders means that the pressure placed on fresh water resources will grow inexorably. About one billion people lack access to safe drinking water, and this number is likely to grow to nearly three billion by 2050 (Gleick 2012). In places that are conflict–prone and vulnerable to water shortages, such as the Middle East, climate change could seriously affect regional stability (Femina and Werrell 2014).

To further complicate this problem, 60% of the world's population lives in crowded water basins shared by multiple states—many of whom are failing, congenital enemies, or both (Postel and Wolf 2001). This is a compelling problem from a national security perspective because most of the world's largest river systems are shared by multiple states. Thus, the possibility of water wars resonates throughout the contemporary national security literature (Diehl and Gleditsch 2001; Gray 2009; Femina and Werrell 2012; DoD 2014).

The U.S. National Intelligence Council warns that the likelihood of water–related conflict will increase during the coming decades (Conca 2006). Nevertheless, many water scholars dismiss this suggestion as exaggerated, and history appears to support their position. An examination of some 1000 international water–related crises during the past 50 years suggests that two–thirds were resolved by cooperative means. This implies that water disputes are not likely to lead to warfare; rather, states tend to resolve these disputes through economic agreements, technological solutions, and diplomacy (Fagan 2011). However, this study argues that the security landscape has changed profoundly, and the history of cooperative water–conflict resolution is no longer a reliable guide to the future. The acceptance of the relationship between water and conflict is gaining momentum, and a number of experts now acknowledge that water wars are certainly plausible; especially if we persist in denying the seriousness of the water crisis in key regions (Soffer 1999; Pearce 2006; Trondalen 2009; IPCC 2014).

This chapter suggests that continued peaceful resolution of interstate water conflicts is inconsistent with the realities of the emerging national security landscape. First, climate change is already affecting the distribution of water in many critical water basins. Second, the proliferation of failing states has singularly reduced the potential for diplomatic resolution in many regions, and that we can no longer continue to rely on quasi-peaceful means using established diplomatic and international protocols to resolve conflicts. Finally, water is an essential resource; however, since 1950, the renewable supply of water *per capita* has fallen by 58% (Fagan 2011; IPCC 2012).

Water shortages will likely provide a tipping point between war and peace for regions already on the brink of conflict, such as the Middle East. States in this region border highly contested water basins and they are facing chronic water shortages combined with the world's fastest growing population. These factors combine to intensify latent ethnic/religious conflicts and decades of distrust and territorial disputes that persist throughout the region. Here, population growth and climate change are fueling a dangerous nexus of water shortages, political instability, and economic stagnation, which are eroding an already unstable situation. This chapter examines linkages between water resources, the problems of transboundary watersheds, and potential conflict.

2 Water and the Security Landscape

The implications of water supply and demand in the Middle East could be severe and represent a departure from the traditional view of security. The Cold War strategic partition of the world dominated the security landscape following the Second World War. In the Cold War scenario, conflicts were state–centric and essentially resulted from politics and ideology. Today, however, the potential for violent conflict triggered by environmental stress looms over society, which is much different from the traditional Cold War concept of security (Myers 1989). Thus, a major shift has occurred: during the Cold War, divisions were created and alliances formed along ideological lines; but now security officials have begun to pay greater attention to problems arising from intensified competition over essential resources (Butts 1997).

Barnett (2004) developed a national security paradigm that attempted to incorporate emerging post–Cold War dynamics—that is economic competition, environmental stress, and failing states. In Barnett's view, these factors are destabilizing large segments of the world because globalization, and the expansion of the global economy that followed, did not lead to an era of integration and world peace. Rather, the unbalanced nature of economic prosperity generated pervasive instability in much of the developing world. This suggests a major shift in the security landscape because Barnett's view of geopolitics presents a world that is segregated into one that it integrating itself into a so-called *Functioning Core*, and one that is trapped in a *Non-Integrating Gap* (Barnett 2004). The *Non-Integrating Gap* is essentially disconnected from the rest of the world; but more importantly, it is inherently unstable and susceptible to environmentally triggered violence (Galgano and Krakowka 2011). Indeed, U.S. policymakers have long acknowledged the destabilizing imbalance of natural resource supply and demand, and its profound consequences for its security interests (Butts 1997). The Arab oil embargoes of the early 1970s quadrupled gasoline prices and clearly pointed out that the global economy depends on highly concentrated deposits of increasingly scarce resources. Although we understand oil as an instigator of conflict, water poses a different and potentially more difficult dilemma because it is a problem that cannot be easily managed (Klare 2001).

Thus, water is fast becoming one of the seminal environmental security factors of the emergent national security landscape because it is an essential resource for which there is no substitute (Butts 1997). Renewable freshwater is fundamental to human society; however, contemporary water demands are approaching the limits of a finite supply (Hensel and Brochmann 2007). Only 0.036% of the world's supply is renewable freshwater; and by 2025, some three billion people (about 40% of the global population) will live in regions that are unable to provide sufficient freshwater to meet basic human needs. Hence, inequities in freshwater supplies will continue to be a source of friction. In fact, 25% of all water–related disputes during the past 50–years have resulted in some form of hostilities—37 have resulted in military action (Gleick 1998; Postel and Wolf 2001; Kreamer 2012).

One of the principal problems affecting water supply and demand, in a universal sense, is that globalization has reduced the friction of distance and created expectations of economic growth in the developing world, and this has escalated the relative disparity between developed and developing states (Butts 2011). In the context of environmental stress and resource competition, the crux of the matter is that global economic output quadrupled after 1950 and during that period, population grew by three billion. However, the problem that looms is that we expect global population to approach nine billion by 2050, and to keep pace, economic output will have to quintuple, which will place greater demands on global freshwater resources (Homer-Dixon 1999). Consequently, water may become an environmental tipping point that triggers violent conflict as greater economic aspirations and human population accelerates demands on the freshwater supply, while at the same time climate change makes supply more uncertain (Gleick 1993).

Contemporary statistics suggest that global water demand for irrigation, domestic, and industrial use will increase faster than the rate of population growth (Fagan 2011). Furthermore, freshwater supplies are, geographically, highly variable and are not equitably distributed in a spatial sense; nor does its spatial distribution match population distribution. The water scarcity problem is further complicated because water does not lend itself to international trade and it is not practical to transport from surplus areas to places of acute scarcity. Water supply is often diminished by water quality issues because increasing populations require more irrigation and dams, both of which can adversely affect water quality. Thus, water passed to downstream users, even in water–rich regions, is typically contaminated (Butts 1997; Pearce 2006).

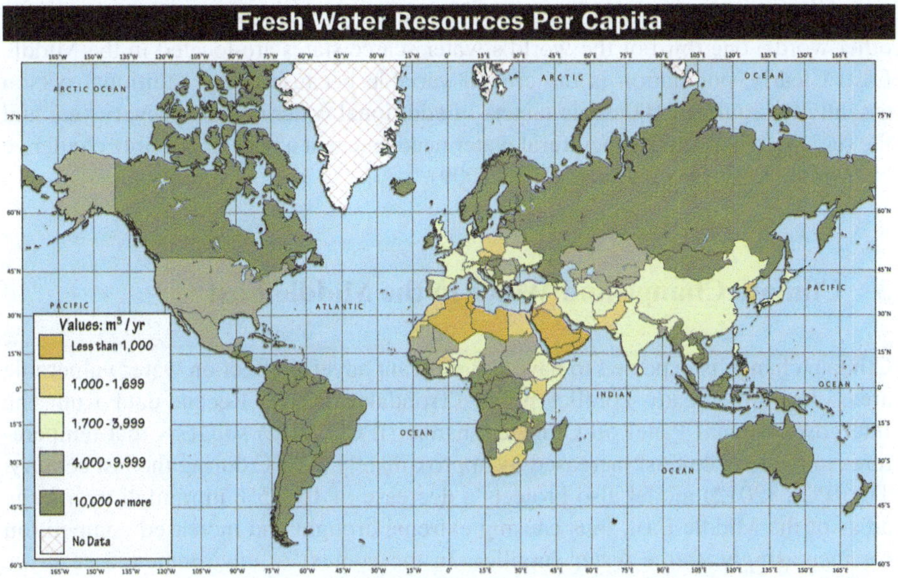

Fig. 1 Water scarcity and stress. After Gleick (1993). Cartography by the author

The geopolitical implications of water supply are challenging as well. From a strategic perspective, upstream states have an advantage in the control of water; downstream states generally remain vulnerable to the political decisions of those upstream. This presents a vexing geopolitical dilemma. According to the United Nations (U.N.), there are roughly 214 international rivers, 150 are shared by two states, and the remainder are shared by three to ten (U.N. 2009). It does not help that coincidently, many of these same river systems are located in the most unstable places on the planet, and unfortunately, in the international legal setting, water law is not effective in settling conflict (Gleick 1993; Soffer 1999). Thus, the method of determining sovereignty over transboundary rivers remains contentious and in the context of places with increased population pressure and severe water stress, the potential for conflict is high (Butts 1997).

Demographically driven increases in water demand during the past 50–years have been unprecedented and forced the Middle East into an acute water deficit. To put this into practical terms, on an annual basis, each individual needs about a cubic meter of water for consumption, about 100 cubic meters for other personal needs, and 1000 cubic meters to grow food (Darwish 1994). Thus, the annual minimum basic need is about 1100 cubic meters: a country with less than 1700 m^3 *per capita* is considered to be water stressed, while states with less than 1000 m^3 *per capita* are considered water scarce (SIWI 2009). Figure 1 illustrates the geographic distribution of water stressed and water scarce states and it confirms the chronic problem that exists in the Middle East. The U.N. (2017) considers 13 states to be water scarce, and four of them—Israel, the Palestinian Authority, Saudi Arabia, and Jordan—are located in this region. U.N. projections suggest that another 10 states will be added to this list by 2025, to include Egypt, Ethiopia, Iran, and Syria: in other words, one third of the world's water scarce states are located in the Middle East. Clearly, population is the crucial variable because every additional person essentially requires 1100 cubic meters of additional water resources every year and the region's associated agricultural water needs present a near impossible challenge for Middle Eastern economies (U.N. 2009).

3 Climate Change and Water in the Middle East

Climate change is expected to have a significant adverse effect on water vulnerable areas that are already conflict–prone (Trondalen 2009). Recent data from the Intergovernmental Panel on Climate Change (IPCC 2012) suggests that temperatures in the Middle East increased by approximately 2–3 °C during the last century. The IPCC (2012) model also projects a decrease of 15–25% in rainfall over large areas of the Middle East, thus causing extreme drought and increased competition for increasingly scarce water supplies. In the context of environmental security, water has two important characteristics that make it a potential source of inter-state warfare: (1) its degree of scarcity, which is being affected in the Middle East by climate change and demographic factors; and (2) the degree to which a water basin is shared between multiple states; a problem that is being exacerbated by poor governance (Gleick 1993; Smith and Vivekananda 2007).

The implications of the various climate models are clear: diminished rainfall, less surface water, lower soil moisture, reduce aquifer recharge, and higher water demand for crops and humans. Yet, it is important to note that there is a degree of spatial variation in temperature and precipitation in the region, and the seasonality of rainfall makes models and generalizations for the Middle East somewhat speculative. Although there is uniformity among temperature predictions, those of precipitation are inconsistent (Trondalen 2009). This is because standard resolution climate models have difficulty representing precipitation in the Middle East, which is modified by complicated topography, inland bodies of water, and proximity of the Mediterranean Sea (Black et al. 2010). While there is variability between climate models, it is nevertheless clear that the Middle East is the world's most water–stressed region, and all projections suggest that climate change will play a role in significantly reducing water availably (Krichak and Alpert 2005, Sappenfield 2007, Black et al. 2010, IPCC 2012).

An analysis of relevant climate models suggests that water vulnerability in the Middle East could be severe. The IPCC (2008) analysis outlines a series of general climate–related problems that are relevant to the Middle East, all of which predict the region will be subjected to prolonged drought and extreme water deficits in the coming decades. Zhang et al. (2005) suggest that the overarching driver of this problem is the statistically significant and spatially coherent trends that suggest a considerable elevation of temperatures throughout the region. However, Zhang et al. (2005) indicate that trends in precipitation indices are less consistent.

Notwithstanding this variability, models suggest statistically significant reductions in rainfall through the end of the century (Trondalen 2009).

Alpert et al. (2008) examined a series of climate models for the Middle East. Their analysis indicates that temperatures in the region increased by 1.5–4 °C during the past 100–years, and that regional temperatures are expected to increase by 4–6 °C by 2100. Simultaneously, precipitation data from the region indicate a dominant negative trend since 1950 (Alpert et al. 2008). Ragab and Prudhomme (2000) developed a monthly climate model for the Middle East, which was developed to predict changes in rainfall from contemporary monthly mean values. Their model suggests that by 2050, most of the region will experience reduced rainfall amounts up to 20–25% lower.

Ragab and Prudhomme (2000) developed a water exploitation index based on the results of their model, which predicts subsequent water demand as a percentage of renewable annual water resources. Their analysis paints a dire picture for the Middle East. For example in the Jordan Valley, it is expected that Israel will exploit 140% and the West Bank/Gaza 169% of the renewable water supply by the end of the century. Kitoh et al. (2008) completed an analysis of climate models that suggests a decrease in the Jordan River's annual flow by as much as 73%. Finally, Weib et al. (2007) suggest that droughts will be 10 times more frequent along with a 15% decrease in rainfall over the next 100 years.

4 The Governance Factor

Contemporary research suggests that countries that are poorly governed and have ineffective institutions will be hardest hit by climate change and water shortages because they lack the capacity to adapt to the adverse effects of climate change and resource shortages (IPCC 2014). Water scarcity will clearly intensify the strain under which those societies already exist. However, environmental stress alone does not, inevitably, trigger warfare. Evidence suggests that it enables or intensifies violent conflict when it combines with weak governance and social division, along with economic inequities to affect a spiral of violence, typically along ethnic and political divisions (Galgano 2007). Contemporary trends indicate that environmentally driven violence has been concentrated in the developing world because it exhibits extreme social fragmentation and stratification (Homer-Dixon 1999). Developing states, like those of the Middle East, are more susceptible to environmentally triggered conflict because they are, characteristically, more dependent on the environment for their economic productivity; and lack the resiliency to overcome these challenges because they have weak economies and small capital reserves, shortages of scientists and engineers, and poor distribution infrastructure (Galgano 2007).

Political instability and weak governance make it problematical for some states to adapt to the physical effects of climate change and water scarcity. Regrettably, governance is a considerable and emergent crisis in the developing world, and since 1990, the number of failing states has grown. Figure 2 illustrates the extent of the

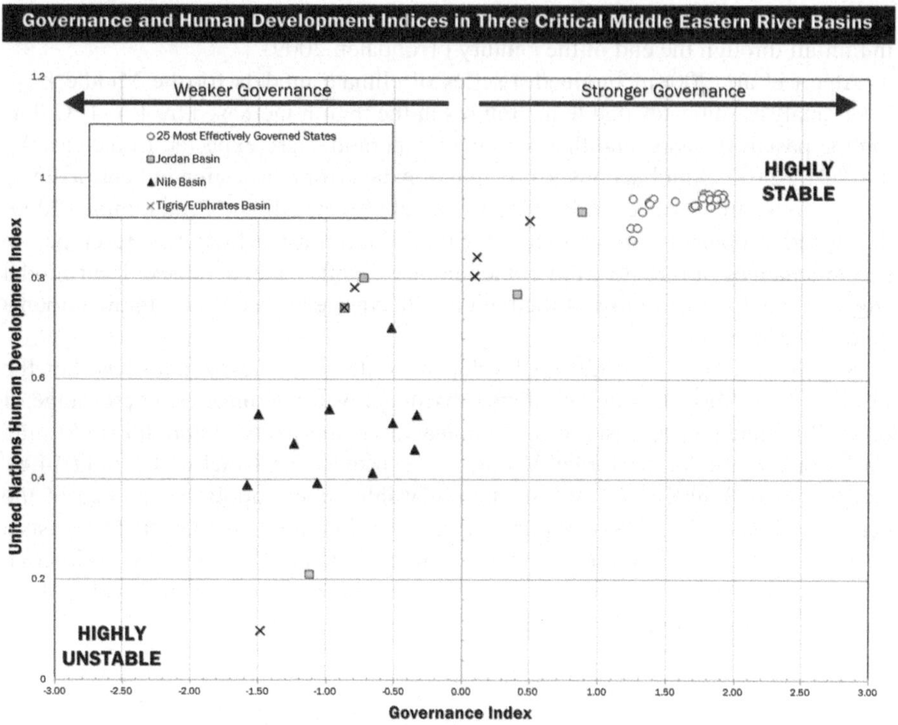

Fig. 2 Governance and development in Middle Eastern river basins. The graph indicates development levels and governance indicators of states within the regions three large river basins. (Sources: U.N. 2015; WGI 2016)

governance problem within the three key river basins in the Middle East. This graph uses the U.N. Human Development Index (HDI) (U.N. 2015) and the World Bank Combined Governance Index (WBGI) to portray the level of stability and government effectiveness in these countries as compared to the world's 25 most effectively governed states. The governance index (WGI 2016) assigns positive and negative values to states. More positive values suggest more effective governance and negative values equate to failing states. The U.N. (2015) developed the HDI as a means of measuring development by merging indicators of human health, education, economic development, and quality of governance into a composite index that can serve as a frame of reference for the overall level of development within a state. The HDI sets a minimum and a maximum with zero equating to low development and one equating to very high development (U.N. 2015). Thus, the data given Fig. 2 suggest that with minor exceptions, most of the states within the Middle East's three principal river basins exhibit low development (i.e., HDI score < 0.5) and weak governance (i.e., WBGI score < 0.0). In fact, the data in Fig. 2 suggest that most are failing states.

Failing states are troubling in this scenario because they have large areas that are outside of effective government control and thus, can be affected severely by humanitarian disasters, environmental stress, and ethnic conflict (Galgano 2007). History indicates that water conflicts were typically resolved in a peaceful fashion. In many cases, leaders of government and non–governmental organizations were able to remedy the problems presented by contested water resources using conventional diplomatic and international protocols. In such cases, well-established doctrine is reasonable as it imparts normative guidance (Rosenthal 2004). However, those same well–established diplomatic protocols and international doctrines and the principles they engender must be now considered in light of the new national security landscape, which is being significantly altered by the double edged problem of climate change and failing states.

5 Water Supply, Demand, and Potential Conflict in the Middle East

Dynamics between population, environmental stress, and conflict are complex, but are not a deterministic recipe. The outcome of a potential environmental security scenario is influenced strongly by government policy, social structure, technology, and infrastructure. The insufficiency of water has in the past led to conflict, and it is currently a critical source of tension in the Middle East. However, we should not assume that water shortage would inevitably lead to war (Amery 2002). Technology, diplomacy, and policy changes can potentially alter the apparent prescription for conflict. However, given rapid population growth, changes in climate, and the imbalance of water, combined with regional political instability, it is reasonable to assume that conflict is a possible outcome. The real problem is that in the Middle East, like the rest of the developing world, the capacity to adapt is declining (Hensel and Brochmann 2007).

The Middle East (Fig. 3) includes three large, transboundary river basins (i.e., Jordan, Nile, and Tigris-Euphrates). The demand for water placed on these three river basins is dictated largely by the region's population, and projections leave little room for optimism. The Middle East manifests some of the fastest growing and urbanizing population in the world, and it also a region within which the withdrawal of water resources are among the highest (Fig. 3), while the renewal rate is the slowest (SIWI 2009). Hence, Middle Eastern states are worrying today about how they will provide drinking water for the extra millions born each year, not to mention agriculture, the primary cause of depleting water resources in the region. In raw population numbers (Table 1), the region is expected to exceed 700 million by 2050—an increase of some 65% during the next 40 years (PRB 2016). The data also indicate that the global rate of natural increase is 1.2% with a mean of 1.7%. However, by comparison, rates of natural increase for the Middle East are striking: the mean rate for the region is 2.23%. Even more noteworthy are the doubling times

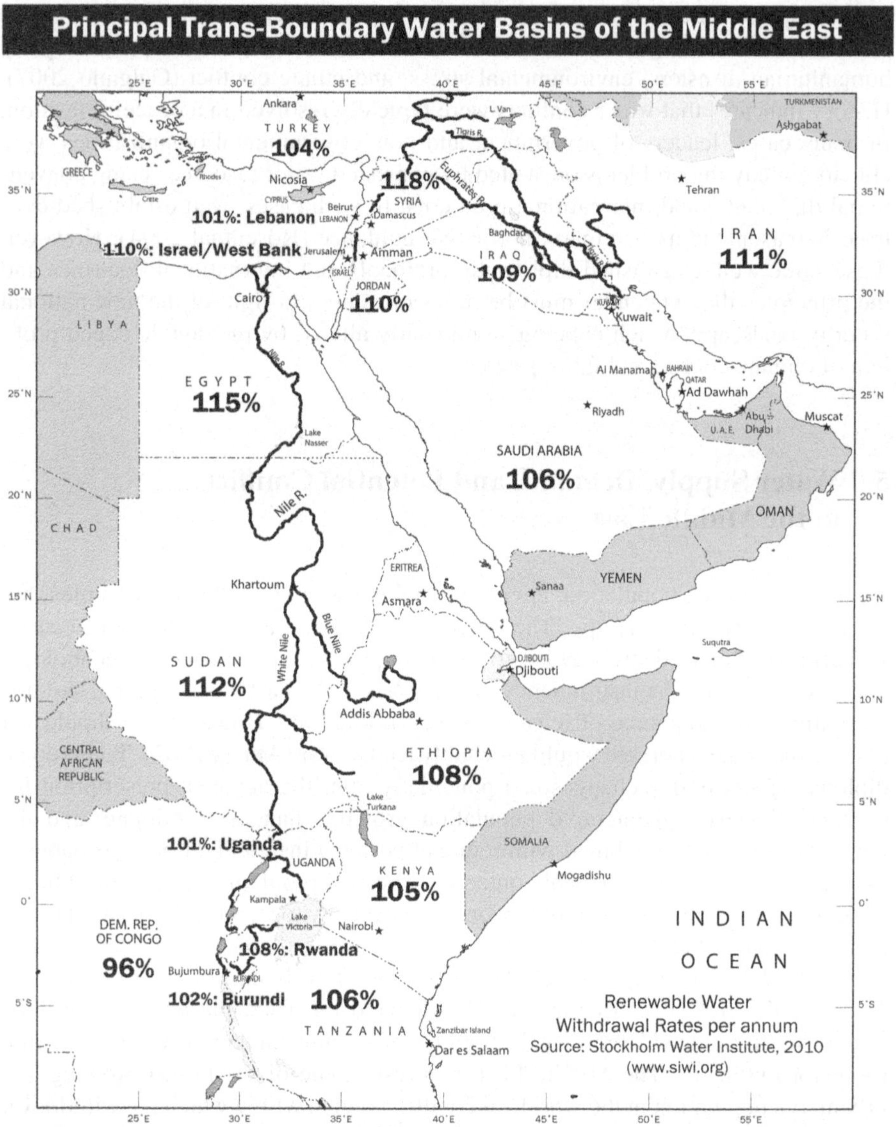

Fig. 3 Map of the three major river basins in the Middle East. The map also indicates the percentage of annual withdrawals of renewable water resources for each state. (Source: Aquastat 2017. Cartography by the author)

illustrated for each state, and the region's doubling time of 31 years, which is nearly half the global rate (PRB 2016). These data (Table 1) are problematical, and given projected changes in climate, and the imbalance of water supply and demand, water continues to be a source of tension. Across the region, *per capita* availability is

Table 1 Comparison of principal states adjacent to major transboundary, Middle Eastern Rivers to global population trends

State	Rate of Natural Increase (%)	Doubling Time (yrs.)	2010 Population (millions)	2025 Population (millions)	2050 Population (millions)
Egypt	2.10	33.0	80.4	103.6	137.7
Ethiopia	2.70	25.0	85.0	119.8	173.8
Iran	2.60	2.60	75.1	87.1	97.0
Iraq	2.60	26.0	31.5	44.7	64.0
Israel	1.60	43.0	7.60	9.40	11.5
Jordan	2.60	27.0	6.50	8.50	11.80
Kuwait	2.00	35.0	3.10	4.10	5.40
Lebanon	1.50	47.0	4.30	4.70	5.40
West Bank/ Gaza	2.80	25.0	4.00	6.00	9.40
Saudi Arabia	2.60	27.0	29.2	35.7	49.8
Sudan	2.20	32.0	43.2	56.7	75.9
Syria	2.50	28.0	22.5	28.6	36.9
Turkey	1.20	58.0	73.6	85.0	94.7
Regional Mean/Total	2.25	31.0	466.0	593.9	772.8
Summary					
Global	1.20	58.0	6829	8108	9485
MDS[a]	0.20	350	1237	1290	1326
LDS[b]	1.70	41.0	4318	5343	6722

Notes:
[a]More Developed States (see PRB 2016)
[b]Less Developed States (see PRB 2016)

already the lowest in the world; and Middle Eastern states are already exploiting nearly all of their renewable water resources (Amery 2002).

As a representative example, the level of the problem for the Jordan River basin is given in Fig. 4. The data illustrate the precipitous decline of water resources and suggest that by 2010, four of the five states in the basin were experiencing severe water scarcity and could not meet the needs of their population from internal, renewable water resources. By 2025, all of the states will be experiencing water scarcity (Aquastat 2017; SIWI 2009), and considering the anticipated increases in the region's population (Table 1), potential scenarios are not encouraging given the region's pervasive instability. The data for the Jordan River basin are particularly discouraging given the highly contested nature of the region's space and the ongoing hostilities that already exist (Gleick 1993).

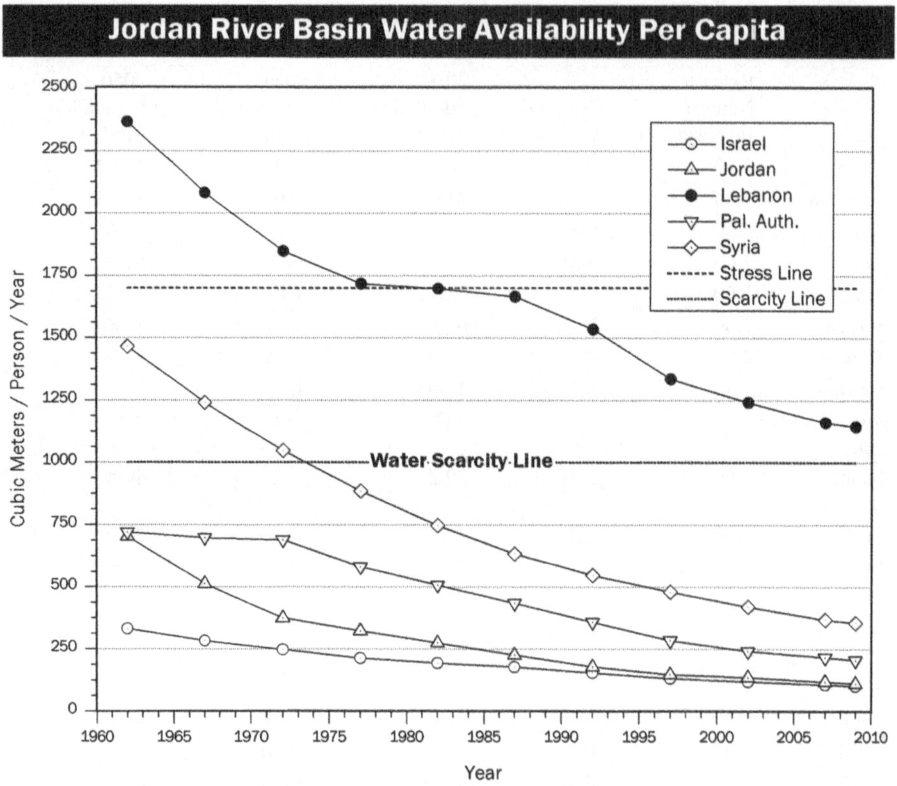

Fig. 4 The extent of the water shortage crisis in the Jordan Valley is illustrated on this chart. These data suggest that most of the states in the basin, with the exception of Lebanon, are below the water scarcity line. This line is based on the idea that each person requires about 1000 m^3 of water each year. (Source: Aquastat 2017)

6 Discussion

Oil has always been the expected trigger for warfare in the Middle East. However, water is now a critical variable because these states depend on three great river systems, or vast underground aquifers, which are already being exploited beyond a sustainable level (Amery 2002). Water has already played an intrinsic role in fostering conflict, altering policies, and changing alliances in the region. During the 1960s, cross border raids against water–related infrastructure were common between Israel, Syria, and Jordan culminating in the 1967 Six-Day War. During the 1964 Arab summit in Amman, Jordan, it was decided to redirect the headwaters of the Jordan River, thus depriving Israel of its most important surface water supply. General Ariel Sharon placed this into context with this quote, "*People generally regard 5 June 1967 as the day the Six-day war began,*" he said. "*That is the official date. But, in reality, it started two-and-a-half years earlier, on the day Israel decided to act against the diversion of the Jordan River*." (Darwish 1994 p. 3).

More recently, Turkey seized an opportunity to exhibit its ability to control the flow of the two great rivers of the Fertile Crescent that emanate from its hinterland. In January 1990, it stopped the flow of the Euphrates. Officially, the disruption was needed to fill the massive lake in front of the new Ataturk Dam; in fact, it was a demonstration to Syria of what might happen if it continued aiding Kurdish rebels in southeast Anatolia (Soffer 1999). Halting the flow of the Euphrates into Syria also brought water shortages in Iraq as well, thus bringing about an alliance between two bitter enemies (Darwish 1994). More importantly, however, Turkey's actions during this episode demonstrate the strategic advantage of an upstream state within a transboundary watershed, and the potential for such activity instigate full–scale military confrontation.

There are short–term solutions to mitigate the effects of water scarcity, such as food imports (e.g., virtual water), desalinization, and international water law practices. However, these are not a panacea. Remarkably, Middle East governments have been able to forestall the seemingly predestined consequences of their escalating water deficits. In the 1970s, water demands in the Middle East could be met from within the region. However, population growth has forced the region into an acute water deficit; and yet, surprisingly, there has been no water war since 1967. Many think that the answer lies in so-called *virtual water*, which is the water contained in imported food (Allen 1998). In fact, more water flows into the region annually as virtual water than flows along the Nile (Darwish 1994). Virtual water has enabled the region to augment its water resources with grain imports and devote scarce water resources to domestic use rather than irrigation, which has reduced tensions and raised the threshold for conflict. However, it is not an enduring solution because virtual water is heavily subsidized. Given the status of the global economy and the fact that droughts are producing food shortages worldwide, continued reliance on virtual water is on shaky ground (Allen 1998). Today, water scarce states account for 26% of grain imports, yet as an additional billion people are added to these water–stressed basins during the next 15 years, and more states join the ranks of food importers, the demand for international grain will exceed supply, thus unbalancing the virtual water flow into the Middle East (Postel and Wolf 2001).

Desalinization is often presented as a popular solution to chronic water shortages and it is being used extensively in localized situations. Nevertheless, desalinization is enormously expensive and cannot meet long–term water demands in the Middle East (Amery 2002). In 2005, more than 13 million cubic meters of fresh water were produced from desalinization each day; nonetheless this represents just under one hundredth of fresh water consumption per day in the Middle East (Conca 2006). Desalination can only be viewed as a short–term solution to resolve or mitigate a very localized water shortage scenario (Gleick 1993).

If warfare over water is to be avoided, steps must be taken to enable an equitable distribution of water in a basin and permit a fair resolution of conflicts (Soffer 1999). International agreements and treaties are certainly desirable, but international law is not very robust. Water law in the U.S. and other parts of the world is well developed and backed by many precedents, and thus conflict resolution can typically rely on well–established doctrine (Butts 1997). For example, in many regions, the legal distribution of water is based on *riparian rights*. This doctrine works well in places where there is a considerable renewable water supply. However, in arid regions, *appropriations doctrine* is more accepted, and under this doctrine, priority is given to the first user of the water (Darwish 1994).

Other doctrines may be appropriate for transboundary water basins. For example, the principle of *equitable apportionment* has been initiated in some regions and appears to be a practicable solution in the Middle East as well. Equitable apportionment calls for a sharing of water benefits equally among states in the watershed on a sustainable basis, regardless of claims to sovereignty. Fortuitously, equitable apportionment is congruent with *Shari'a Law*. For example, under this religious principle people who dig a well maintain rights to first use, but cannot refuse its use for drinking to others. Thus, a state would have full possession of only the amount of water it needs at a precise moment, thus leaving sufficient water for its downstream neighbors (Darwish 1994).

7 Summary and Conclusions

International law is, unfortunately, not robust or clear in settling water conflicts. The alternative doctrine of equitable apportionment is attractive, but predictably, in the context of the Anatolia Dam water conflict, Turkey maintains the position that it has complete sovereignty over the basin because it is the upstream state. However, Iraq and Syria clearly support the doctrine of equitable apportionment, insisting on a reasonable distribution of water based on need and historical use. Noticeably absent, however, and a guarantee that international water law will remain inadequate, is an enforcement mechanism.

In the past, the resolution of water–related conflict was typically achieved through diplomacy, economic cooperation, and technology. However, the emerging security landscape is far more complex and is being affected by the growing inequities between developed and developing states, and the pervasiveness of failing states. Globalization of the economy and population growth, combined with greater expectations of increased economic affluence, are going to place greater demands on resources and exacerbate the problems of resource supply and demand—certainly the world's renewable freshwater resources will be strained beyond sustainable levels and may become a tipping point for violent conflict. This would be difficult enough, however the security situation will be intensified by climate change, which may dramatically reduce rainfall and river discharges in the world's most populated and water-stressed regions.

Water resource scarcity is an environmental security issue that currently exercises considerable influence on regional stability. Projected trends in population growth, water demand, and climate change could make water scarcity far more prominent on the geopolitical and security landscape. Although the role of water as a possible trigger for violent conflict on an inter-state scale remains a hypothetical exercise, water issues will continue to be a strategically important variable on the national security landscape, and they should be used as an indicator of impending regional instability and a persistent reminder of the significance of geographical variables to military security affairs.

References Cited

Allen T (1998) Watersheds and problemsheds: explaining the absence of armed conflict over water in the Middle East. Middle East Rev Int Aff 2(1):49–51
Alpert PI, Krichak KO, Shafir H, Haim D, Osetinsky I (2008) Climatic trends to extremes employing regional modeling and statistical interpretation over the E. Mediterranean. Global Planet Change 63:163–170
Amery HA (2002) Water wars in the Middle East. Geogr J 168(4):313–323
Aquastat (2017) Main country database. The Food and Agricultural Organization of the United Nations. http://www.fao.org/nr/water/aquastat/dbases/index.stm. Accessed May 2017.
Barnett T (2004) The Pentagon's new map. G. P. Putnam and Sons, New York
Black E, Brayshaw DJ, Rambeau MC (2010) Past, present, and future precipitation in the Middle East: insights from models and observations. Philos Trans R Soc 368(1931):5173–5184
Butts KH (1997) The strategic importance of water. Parameters 27(2):65–83
Butts KH (2011) Environmental security: a growing force in regional stability. In: Galgano FA, Palka EJ (eds) Modern military geography. Routledge, New York, pp 54–64
Conca K (2006) Governing water, Cambridge. Massachusetts Institute of Technology Press
Darwish A (1994) Water wars: the next major conflict in the Middle East. Middle East News. http://www.mideastnews.com/WaterWars.htm. Accessed Jan 2012.
Department of Defense (DoD) (2014) Quadrennial defense review. U.S. Department of Defense, U.S. Government Printing Office, Washington, DC http://www.defense.gov/pubs/2014_Quadrennial_Defense_Review.pdf. Accessed 6 May 2017
Diehl PF, Gleditsch NP (eds) (2001) Environmental conflict. Westview Press, Boulder
Fagan B (2011) Fresh water supplies are going to run out, so what can we do to make the taps keep running? The Independent, 30 June 2011. http://www.independent.co.uk/environment/nature/fresh-water-supplies-are-going-to-run-out-so-what-can-we-do-to-make-the-taps-keep-running-2304481.html. Accessed 30 Apr 2012.
Femina F, Werrell CE (2012) Update: climate and security 101: Why the U.S. national security establishment takes climate change seriously. The Center for Climate and Security, Briefer no. 23, 26 February 2014. https://climateandsecurity.org/reports/#BRIEFERS. Accessed 12 Jan 2017
Femina F, Werrell CE (2014) Climate change, drought, and social unrest. The Center for Climate and Security, Briefer no. 11. https://climateandsecurity.org/reports/#BRIEFERS. Accessed 12 Jan 2017
Galgano FA (2007) A geographic analysis of ungoverned spaces. PA Geogr 44(2):67–90
Galgano FA, Krakowka A (2011) The environment's influence on regional stability and conflict. PA Geogr 48(2):3–32
Gleick PH (1993) Water and conflict. Int Secur 18(1):79–112
Gleick PH (1998) The world's water, the biennial report on freshwater resources. Island Press, Washington, DC
Gleick PH (2012) The world's water, volume 7, the biennial report on freshwater resources. Island Press, Washington, DC
Gray CS (2009) National security dilemmas: challenges and opportunities. Potomac Books, Inc, Washington, DC

Hensel PR, Brochmann M (2007) Armed conflict over international rivers: the onset of militarization of river claims. http://www.paulhensel.org/Research/apsa08r.pdf. Accessed Apr 2012.
Homer-Dixon TF (1999) Environmental scarcity and violence. In: Princeton. Princeton University Press
Intergovernmental Panel on Climate Change (IPCC) (2008) Technical paper on climate change and water. IPCC technical paper VI. http://www.ipcc-wg2.org/. Accessed May 2012.
Intergovernmental Panel on Climate Change (IPCC) (2012) Managing risks of extreme events and disasters to advance climate change adaptation. Special Report of Working Groups I and II of the Intergovernmental Panel on Climate Change, eds. Field CB, Barros V, Stocker TF, Qin D, Dokken DJ, Ebi KL, Mastrandrea MD, Mach KJ, Platter GK,. Allen SK, Tignor M, Midgley PM. Cambridge University Press, Cambridge, UK. http://ipcc-wg2.gov/SREX/. Accessed 21 July 2013
Intergovernmental Panel on Climate Change (IPCC) (2014) Climate change 2014: impacts, adaptation, and vulnerability. Summary for policy makers. Working Group II Contribution to the Fifth Assessment Report of the Intergovernmental Panel on Climate Change. Cambridge University Press, Cambridge, UK. http://www.ipcc.ch/report/ar5/wg2/. Accessed 8 Nov 2016
Kaplan RD (2000) The coming anarchy: shattering the dreams of the post cold war, New York. Random House
Kitoh A, Yatagai A, Alpert P (2008) First super-high resolution model prediction that the ancient "Fertile Crescent" will disappear this century. Hydrol Res Lett 2:1–4
Klare MT (2001) The new geography of conflict. Foreign Aff 80(3):49–61
Kreamer DK (2012) The past, present, and future of water conflict and international security. J Contemp Water Res Educ 149(1):87–95
Krichak SO, Alpert P (2005) Decadal trends in the East Atlantic – West Russian pattern and Mediterranean precipitation. Int J Climatol 25(2):183–192
Myers N (1989) Environment and security. Foreign Policy 74:23–41
Pearce F (2006) When the rivers run dry. Beacon Press, Boston
Population Reference Bureau (PRB) (2016) 2016 world population data sheet. http://www/prb.org. Accessed Jan 2017.
Postel SL, Wolf AT (2001) Dehydrating conflict. Foreign Policy, 18 September 2001
Ragab R, Prudhomme C (2000) Climate change and water resources management in the southern Mediterranean and Middle East countries. The Second World Water Forum, The Hague, pp 17–22
Rosenthal J (2004) New rules for war? Naval War Coll Rev 58(3/4):91–101
Sappenfield M (2007) Six places in the world where climate change could cause political turmoil. The Christian Science Monitor, 6 December 2007, p 14
Smith D, Vivekananda J (2007) A climate of conflict. International Alert, London http://www.international-alert.org. Accessed May 2012
Soffer A (1999) Rivers of fire: the conflict over water in the Middle East. Rowman and Littlefield, London
Stockholm International Water Institute (SIWI) (2009) Water resources in the Middle East. http://siwi.org. Accessed Apr 2012.
Trondalen JM (2009) Climate changes, water security, and possible remedies for the Middle East. United Nations World Water Assessment Programme, Scientific Paper. United Nations Educational, Scientific and Cultural Organization, Paris
United Nations (U.N.) (2009) Water in a changing world, United Nations world water development report 3. http://publishing.unesco.org/. Accessed May 2012
United Nations (U.N.) (2015) Human development report 2015. Technical notes. http://hdr.undp.org/sites/default/files/hdr2015_technical_notes.pdf. Accessed 1 Feb 2017
United Nations (UN) (2017) Wastewater, the untapped resource. United Nations World Water Development Report 2017. United Nations Educational, Scientific and Cultural Organization, Paris. http://www.unesco.org/new/en/natural-sciences/environment/water/wwap/wwdr/2017-wastewater-the-untapped-resource. Accessed 2 Dec 2017.

Weib M, Florke M, Menzel L, Alcamo J (2007) Model based scenarios of Mediterranean droughts. Adv Geosci 12:145–151

Worldwide Governance Indicators (WGI) (2016) The worldwide governance indicators (WGI) dataset. The World Bank. http://info.worldbank.org/governance/wgi/index.aspx#home. Accessed 28 Dec 2016.

Zhang XB, Aguilar E, Sensoy S, Tonyan H, Tagiyeva U, Ahmed N, Kutaladze N, Rahimzadeh F, Taghipour A, Hantosh TH, Albert P, Semawi M, Ali MK, MHS A-S, Al-Oulan Z, Zatari T, Khelet IA, Hamoud S, Sagir R, Demircan M, Eken M, Adiguzel M, Alexander L, Peterson TC, Wallis T (2005) Trends in Middle East climate extreme indices from 1950 to 2003. J Geophys Res Atmos 110:213–256

Water, Land, and Governance: Environmental Security in Dense Urban Areas in Sub-Saharan Africa

Amy K. Richmond

Abstract As the demographic shift to an urban society continues, there are often extreme limitations in the abilities of formal municipal governments to plan space, infrastructure, and resources. This leads to issues of large-scale unplanned habitations, extreme stress on environmental resources, uncontrolled sprawl, pollution, informal governance structures, and dangerous power conflicts. This chapter will use specific examples from Accra, Ghana and Kampala, Uganda to discuss Environmental Security in sub-Saharan Africa. Both these cities are mid-size cities experiencing rapid urbanization and both cities have a significant proportion of their population living in informal areas. Environmental security in dense urban areas in sub-Saharan Africa is intricately linked to the complex relationship between water, land, and governance.

Keywords Africa · City · Governance · Health · Informal area · Land tenure · Megacity · Non-Governmental Organization · Poverty · Sanitation · Slum · Urban · Water

1 Dense Urban Environments and Environmental Security

The megacity has quickly become a symbol of the twenty-first century human environment, with over 31 megacities in existence across the world in 2016 (United Nations 2016). The rise of the megacity has brought forth significant issues in planning, authority, and stability as these rapidly expanding urban centers outpace the capacities of their government structure. With populations in excess of 10 million, there are often extreme limitations in the abilities of formal municipal governments to plan space, infrastructure, and resources. This leads to issues of large–scale unplanned habitations, extreme stress on environmental resources, uncontrolled sprawl, pollution, informal governance structures, and dangerous power conflicts.

A. K. Richmond (✉)
Department of Geography and Environmental Engineering, United States Military Academy, West Point, New York, USA
e-mail: amy.richmond@usma.edu

F. Galgano (ed.), *The Environment-Conflict Nexus*, Advances in Military Geosciences, https://doi.org/10.1007/978-3-319-90975-2_6

As this demographic shift to an urban society continues, it becomes increasingly important to understand this new terrain. Massive growth creates increasing social conflict and environmental strain (Richmond et al. 2016a).

Over 25 of the world's 100 fastest growing cities are in Africa, even though Africa's population is still below the 50% urban threshold (UN HABITAT State of the World Cities 2016). Sub–Saharan Africa is the world's least urbanized, yet fastest urbanizing region. By the end of the current decade the urban population in Africa is expected to increase by 50% (UN HABITAT State of the World Cities 2016). The impacts of continued rapid growth in African cities include acute housing shortages, traffic congestion, pollution and uncontrolled peri–urban sprawl, and high levels of poverty and inequality.

The risk is that Africa's informally controlled urban areas will detach themselves even further from effective government control, and instead be made up of local, and possibly fractured or criminal power structures that may constitute a threat to the population of the city and the state. Africa's mid and small cities will account for 75% of its urban growth (UNDP MDG Report 2014). These cities usually have fewer resources and less planning to deal with this growth than megacities. Many of these informal areas are also the most susceptible to disasters because they are located on marginal lands and consequently will be the main target of relief efforts. Lack of adequate emergency and relief services place these populations at a great risk in the event of a disaster. Focus needs to be given to effective urban planning, which must include the sustainable management of environmental resources.

Despite the urban surroundings, people in the majority of African cities still do many of the daily tasks, such as fetching water, that they did in rural environments. Perhaps the two most influential environmental variables in dense urban environments are water and land. The procurement of both can be tenuous and can create friction points. Access to land and water is usually controlled by informal governance structures that fill the void created by an ineffective state government. Informal governance can range from efficient, transparent, and equitable to criminal groups controlling access and price gouging. As the 2014 State of African Cities report concluded, "*African cities are not fulfilling their development potential, which is underscored by evidence of increasing environmental strains and social conflict in urban areas*" (New York and Washington 2015).

This chapter will use specific examples from Accra in Ghana and Kampala in Uganda to examine environmental security in sub–Saharan Africa. Both cities are mid–sized and are experiencing rapid urbanization; and both have a significant proportion of their population living in informal areas. Water is the primary driver of environmental insecurity. However, access to water is coupled to land tenure. Land owners can control the location of water sources, the price of water, and access. Land owners are the community leaders and form the majority of the informal governance structures. Consequently, environmental security in dense urban areas in sub-Saharan Africa is intricately linked to the complex relationship between water, land, and governance.

1.1 What Is Environmental Security in the Context of a Dense Urban Environment?

Environmental security describes the process involving environmental risk analysis based on multifaceted linkages between anthropogenic and natural processes that destabilize the environment and contribute to instability or conflict. The fundamental components of environmental security include: (1) environmental processes that undermine governments and promote instability; and (2) environmental processes that trigger civil conflict (Krakowka et al. 2012). In the context of a dense urban environment there are primarily three components that influence environmental security. First, degraded and environmentally stressed areas outside the city can reduce livelihoods and create a situation where people are forced into congested urban areas in search of a better livelihood. Second, unmanaged and unregulated growth results in a significant strain on environmental resources within the city. This is primarily realized in the vast amounts of pollution that unregulated growth produces, which in turn contaminates water sources, makes air toxic, degrades land, and strains environmental systems such as wetlands. All of this decreases human health and welfare. And finally, the poorest communities are often located on marginal land—such as steep slopes and wetlands—which makes them particularly vulnerable to natural disasters. This raises the complexity of the problem for government leaders as well as directors of non–governmental organizations and intergovernmental bodies as they attempt to develop relief strategies.

1.2 What Is Informality?

Informality describes the absence of government regulation and consequently it is not taxed or regulated by the state. The poor engage in informal activity because their only alternative is to exist outside of official law (Soto 2001). The fundamental motivation for people to engage in informal activity is resource scarcity, which stems from land, food, or service scarcity. Activities and structures arise that allow individuals to cope or thrive in those circumstances. The International Labor Organization (ILO 2009) now estimates that informal work accounts for 72% of all employment in sub–Saharan Africa. The spatial impact has been massive as these populations compete for scarce land near crowded urban cores.

1.3 What Is a Slum or Informal Settlement?

The term informal settlement is nearly interchangeable with the term slum (Myers 2016). The U.N. defines a slum as a contiguous settlement within which the inhabitants are characterized as having inadequate housing and basic services. A slum is usually not recognized by public authorities as an integral part of the city (UN Habitat 2003). The U.N. describes a slum household as a group of individuals living under the same roof lacking one or more of the following conditions: (1) access to improved water; (2) access to improved sanitation; sufficient-living area; (3) durability of housing; and (4) security of tenure (UN Habitat 2003).

Slums are a clear manifestation of poorly planned and mismanaged urban governance (Nyametso 2012). In sub–Saharan African cities, a significant proportion of the population live in slums. For example, 54% of Kampala's 2014 population lived in a slum and in Accra that number is estimated to be 38% (World Bank Group 2017).

2 Water, Land, and Governance

Water and sanitation are a leading cause of environmental insecurity in slums in sub-Saharan Africa. However, it is impossible to focus exclusively on water because its availability and quality is so closely related to land tenure and governance. Understanding land tenure and governance enables a better understanding of why water is a dominate driver of environmental insecurity.

2.1 Water

Many countries in the region exist in a perpetual state of water crises resulting from a variety of large scale and small scale factors. Climate change is a large scale cause and results in shifting rain patterns, loss of tree coverage, and poor soil fertility. Consequently, farmers are typically forced to migrate into cities, adding to already existing population pressures. A small scale factor is the contamination of water sources by ungoverned pollution. For example, water in Ghana is highly contaminated because of illegal gold mining by the Chinese, Ukrainians, Russians, Ghanaians, and other regional actors. Unregulated amounts of mercury, used in the gold mining process, are dumped directly into the Volta river tributaries, which is the main source of Accra's fresh water. Hence, water resources are contaminated and fishing is devastated. The Ghanaian EPA, local Chiefs, and politicians are contributing to this degradation by accepting bribes. In Uganda, uncontrolled pollution in Lake Victoria, Kampala's primary source of water, has decreased the city's water quality significantly. Pollution comes from a variety of other sources that include illegal car washing facilities, heavy industry along the lake, and garbage dumping (Richmond et al. 2016b). At an even smaller scale, individual wells and springs are frequently contaminated by raw sewage from open pit latrines.

Water access and sanitation (i.e., WASH) are key components of environmental insecurity in slums. Without access to clean water, waterborne diseases such as Cholera persist and decrease the overall health and productivity of the community.

Fig. 1 Open spring in Bwaise, Uganda. Pit latrines are located directly behind the spring (not pictured)

Water for urban areas is complicated by the poor location of housing and water sources. Areas of high poverty and dense population face challenges of water availability and affordability. The role of women and children, as the primary water collectors, is arduous and time consuming. They travel considerable distances to purchase water for their homes, which prevents girls of middle school age and older from attending school regularly. Security is also an issue for women who collect water because they are vulnerable to sexual violence as they travel to water sources. Water quality is impacted by limited adequate sanitation facilities and the disposal of most garbage and sewage in open drains. Water is frequently more expensive for people in slums; far higher than the cost in the rural areas.

2.1.1 Water Availability

Water in informal dense urban environments comes from a variety of sources including, open sources, piped water, tanks, wells, and bags. Access depends on a variety of factors including, land tenure, community governance, price, and distance. Open springs (Fig. 1) are free and access is generally uncontrolled. Quality, however, tends to be poor as the vast majority of these springs are highly contaminated, usually because of their close proximity to pit latrines. Piped and groundwater can be expensive. However, it is relatively cleaner than the open springs. However, piped and groundwater usually need to be boiled before use. Both piped and groundwater are commonly known as standpipes and these are managed either by a nearby storeowner or by a kiosk manager who charges for the service. It is highly unusual to have a tap operational in a typical slum home because land ownership is required for this service and the user has to pay a connection fee. Ownership of water sources can range from organized community groups, funded by NGOs, to criminal organizations. It is not uncommon for water in slums to cost considerably more than water outside of the slum. For example, in Kampala, the national government charges about 20–40 shillings (i.e., USD 0.005–0.01) for 20 liters of water. In a slum, the same amount is usually 200–400 shillings (i.e., USD 0.06–0.11). Households that cannot afford to pay for water are forced to use contaminated open springs. In most cases, affordable water in slums usually relies on NGO involvement to help provide it.

It is also common for the city water supplier to provide water to specific locations within a community. In such cases, the community then decides how to manage the tap. People will pay the tap manager who then pays the municipality. Taps, however, are frequently turned off by the city water municipality for hours at a time with no notice, either to fix existing infrastructure or because of water shortages. When this happens, everybody in the community must use the contaminated spring water or walk long distances to a tap that has not been turned off, and endure long wait times.

Boreholes are also present and are usually built by an NGO or other outside organization. The community usually elects a water board that oversees the management of the water. This water is usually higher in quality than water provided by the city; and is more affordable. Community members choose their water source based on location, distance, and affordability.

2.1.2 Water Sanitation

Water quality is severely affected by pollution in dense urban environments and pit latrines are a primary source. Despite frequent attempts by the government to regulate them, pit latrines remain in wide use and are the primary toilet in these areas. In Kampala, during the rainy season, pit latrines typically empty into drainage ditches, which then drain into local water bodies. A general lack of adequate infrastructure exacerbates this situation. Only 7% of the population in Kampala has access to an improved sanitation facility at the household level (UN Habitat 2006). In Accra, that number is 15% (UNICEF 2017). Insufficient sewer infrastructure is expected to persist for decades, as a citywide sewer system not feasible. The majority of planned sanitation takes the form of poor quality septic tanks that frequently leak into groundwater. In Accra, septic tanks are emptied by pump trucks that, until recent laws restricted this, disposed of their contents into the Atlantic Ocean. Recently, private sewage facilities have been built to receive this waste and process it

Fig. 2 Drainage ditch clogged with garbage in Bwaise, Uganda

correctly. However, the demand far exceeds the capacity and so illegal dumping still continues.

Water quality is also influenced by solid waste, which clogs storm water drains, exacerbating flooding and pollution of surface water (Figs. 2 and 3). Garbage also collects in surface water (Fig. 4). The absence of an adequate road network significantly restricts municipal waste collection. Trucks that do make it into the community cannot reach most homes and they do not come at reliable times. However, in the absence of city-wide garbage collection the informal market can often fill in the gap. Local entrepreneurs collect garbage from dwellings for a fee and dispose of the waste. They often travel through the narrow roads by foot or motorcycle and bring garbage to a collection point (Fig. 5).

2.2 *Land Tenure*

Reliable access to water and sanitation relies heavily upon land tenure. To provide adequate water and sewage hookups for urban areas, and to reduce pollution and increase sanitation facilities, the government has a delicate balance of working with

Fig. 3 Drainage ditch clogged with garbage Old Accra, Ghana

the informal sector and navigating the complicated land tenure system that exists in many African cities. Often informal systems of land registration rely on oral traditions or verification of land use by neighbors, which complicates any process and makes it nearly impossible to trace land ownership. Land, and even space in rooms, is usually passed down from generation to generation. Land tenure disputes are a principal source of conflict in informal settlements. These disputes occur at all scales within the community and they inundate the court system.

Land tenure disputes also occur at the macro–scales as well in cases in which the government opposes the existence of an informal community. This condition can persist for a variety of reasons, but usually it is because the slum is located on government–owned land, or on a future development site. Consequently, the government will withhold services in an attempt to dissuade people from occupying the land. The government will frequently threaten eviction, which reduces slum improvements by residents because of the investment risk. This situation is difficult to remedy because any attempts to forcibly clear the slum usually results in violence and eventual resettling on the same land because the residents have nowhere else to go. Even if new housing is provided for the cleared population, it is commonly located on the outskirts of the city far away from their livelihood. Without affordable and reliable transportation, most people end up resettling the area or crowding an existing neighboring communities, thus perpetuating the problem.

Fig. 4 Garbage surrounding Old Accra, leading to a lagoon that goes directly to the Atlantic Ocean a short distance away

Land tenure systems in most sub-Saharan African cities are complex and non-transparent. In Kampala, five different land tenure systems exist, many of them overlapping. Different land types offer different levels of security. For example, public land provides little security because the government can easily clear it. In contrast, Mailo land comes with a title and is rented with 99-year leases from the Crown, making them reasonably secure from evictions. A reliable map of land types is hard to come by and Kampala's city planners rely on topographic maps that date to British colonial days. Kampala's city government often resorts to slum clearing instead of upgrading infrastructure, which increases environmental insecurity because the displaced population crowds into neighboring communities placing further strain on resources. In Accra, all land is owned by the government or by Chiefs. Land can be leased with 99-year leases and these can be sold. It is impossible to keep track of who owns the land or when the leases have expired. As a result, courts and Chiefs spend a considerable amount of their time trying to settle land disputes.

In addition to land tenure issues it is also important to note that most slums are located on environmentally vulnerable and marginal land. This makes them particularly vulnerable to natural disasters such as severe flooding. Such events displace people; and force them to resettle, usually with relatives already living in crowded dwellings.

Fig. 5 Garbage Collection in Nema, Ghana. Garbage collection is an example of an informal business

2.3 Governance: The Control of Land and Water

Governance in sub-Saharan Africa is complicated and is usually a patchwork of traditional chiefs and elected officials. For example, Ghana is governed by elected government officials (i.e., formal) and Chiefs (i.e., informal). Corruption is pervasive in both forms of government; and without enforcement of laws communities will remain environmental insecure.

Access to water and sanitation facilities is dictated predominately by informal governance groups at the community level. At the most basic level, the location of infrastructure depends largely on land ownership and the willingness of the owner to permit the land to be used for his purpose. Frequently, community leadership and landowners determine where to locate infrastructure and how to manage its use. Because people pay for water, or to use a latrine, money is collected and that pays for the maintenance of the facility.

As mentioned above, the cost of water in slums is more expensive. Without adequate oversight, often facilitated by a strong community leadership board and supported by an NGO, corruption can lead to the high price of water. Strong community leadership can take a variety of forms ranging from savings groups, religious leadership, and woman's organizations. The most sustainable and effective leadership groups are diverse and include women and men, different religious leaders, land owners, and community entrepreneurs. They also have the backing and support of an NGO and less frequently the formal government. It is imperative that the formal government, or any other organization that wants to do projects within the community, channel their efforts through these leadership groups. It is these grassroots-level initiatives that contribute the most to increasing welfare and reducing environmental vulnerability.

3 Summary

Sub–Saharan environmental security is largely predicated complicated relationships between water, land tenure, and governance. Access to clean water is essential to human welfare and reducing environmental insecurity. To understand who controls access to water, it is critical to appreciate land tenure, a system that is highly complicated. Landowners wield considerable power in informal communities and form the foundation of most informal governance structures. Governance in slums can range from strong, productive, grassroots organizations that provide clean and functioning sanitation facilities, affordable water, and organize for waste reduction to criminal groups that severely restrict access to resources. For these areas to improve, urban planning is critical and must involve established leadership groups within the community. Examples of slum improvement projects are easy to find and their influence and success is abundantly clear. In stark contrast to these well–organized areas are those where criminal groups control water and land. In these scenarios, community welfare is significantly reduced and the overall quality of life is lower. In criminally run communities there is an excess of pollution and garbage, water is expensive and the quality is poor and not monitored. Often the government has no interest in slum improvement projects because their focus is on slum clearing. These areas remain largely ungoverned and make excellent hiding places for criminals and non-state actors.

The population in informal communities depends on the environment for the resources that it supplies–most significantly water. In turn the community creates large amounts of pollution that compromises livelihood and wellbeing by threatening the quantity and the quality of the environmental resources. Without urban planning, adequate regulation, and the ability to enforce regulations, many dense urban areas are stuck in a cycle that is difficult to immerge from. Breaking this cycle of poverty is challenging but necessary, especially because urbanization is increasing at such rapid rates, pushing people into increasingly vulnerable land, and rapidly increasing pollution levels.

Literature Cited

International Labor Organization (ILO) (2009) The informal economy in Africa promoting transition to formality challenges and strategies. International Labor Organization, Geneva

Krakowka AR, Heimel N, Galgano FA (2012) Modeling environmental security in Sub-Saharan Africa. Geograph Bull 53(1):21–38

Myers GA (2016) Urban environments in Africa: a critical analysis of environmental politics. Policy Press, Great Britain

New York and Washington D.C. (2015) U.N. Department of economic and social affairs. World urbanization prospects the revision human settlements program. Situation analysis of informal settlements in Kampala Nairobi Kenya UNHABITAT United Nations Habitat state of African cities reimagining sustainable urban transitions rep the World Bank February the growth challenge can Ugandan cities get to work World Bank Group 2014 SRC-GoogleScholar

Nyametso JK (2012) Resettlement of slum dwellers, land tenure security and improved housing, living and environmental conditions at Madina estate, Accra, Ghana. Urban Forum 23(3):343–365

Richmond AK, Arney C, Coronges K, Simonson L (2016a) The role of network science in analyzing slums in rapidly growing urban areas. Small Wars J. August. http://smallwarsjournal.com/jrnl/art/the-role-of-network-science-in-analyzing-slums-in-rapidly-growing-urban-areas. Accessed 23 June 2017

Richmond AK, Link J, Malcomb DW (2016b) Case study 1: the impact of pollution in a multi-country lake: a case study of Lake Victoria in the Ugandan context. In: Gaskin-Reyes C (ed) Water planet: the culture, politics, economics, and sustainability of water on Earth: the culture, politics, economics, and sustainability of water on earth. ABC-CLIO, New York

Soto H (2001) The mystery of capital: why capitalism triumphs in the west and fails everywhere else. Black Swan, New York

UN Habitat (2003) The challenge of slums: global report on human settlements 2003. https://unhabitat.org/?mbt_book=the-challenge-of-slums-global-report-on-human-settlements-2003. Accessed 23 June 2017

UN Habitat (2006) Situation analysis of informal settlements in Kampala. http://unhabitat.org/books/situation-analysis-of-informal-settlements-in-kisumu/. Accessed 17 June 2017

UN HABITAT State of the World Cities (2016) http://mirror.unhabitat.org/documents/media_centre/sowcr2006/SOWCR%205.pdf. Accessed 23 June 2017

UNDP MDG Report (2014) Assessing progress in Africa toward the millennium development goals. Analysis of the common African position on the post-2015 development agenda. https://www.google.com/search?q=http://www.undp.org/content/dam/rba/docs/Reports/MDG_Africa_Report_2014_ENG.pdf&ie=utf-8&oe=utf-8. Accessed 23 June 2017

UNICEF (2017) Assessment of waste water treatment in Accra. Civil Engineering Department KNUST. https://www.unicef.org/ghana/assessment_of_waste_water_plant_report.pdf. Accessed 26 September 2017

United Nations (2016) The world's cities in 2016. http://www.un.org/en/development/desa/population/publications/pdf/urbanization/the_worlds_cities_in_2016_data_booklet.pdf. Accessed 23 Sept 2017

World Bank Group (2017) Population living in slums (% of Urban Population) | Data. https://data.worldbank.org/indicator/EN.POP.SLUM.UR.ZS?end=2014&locations=GH&start=2013. Accessed 27 Sept 2017

When Politics, the Environment, and Advocacy Compete–Environmental Security in the South China Sea

Wiley C. Thompson

> *"Victory will be based on the full support of a prosperous and contented population while engaging in strategic diplomacy in step with military preparations."*
>
> Geopolitics and the Dragon's Advance: An Exploration of the Strategy and Reality of China's Growing Economic and Military Power and its Effect Upon Taiwan
>
> *"Environmental scarcity has insidious and cumulative social impacts, such as population movement, economic decline, and weakening of states. These can contribute to diffuse and persistent sub-national violence. The rate and extent of such conflicts will increase as scarcities worsen."*
>
> Environmental Scarcities and Violent Conflict: Evidence from Cases
>
> Dr. Thomas Homer-Dixon

Abstract Island building in the South China Sea by China and other neighbors continues at a destructive and unprecedented rate. China now occupies more than 3000 acres of artificially constructed island space and has built land at a pace that is 17 times greater in recent years than all other claimants have built in combined efforts over the past 40 years. While calls for accountability by some national actors have been insistent, voices from non–governmental actors are largely absent.

The views and opinions expressed in this article are those of the author and do not necessarily reflect the official policy or position of the Department of Defense, US Pacific Command, or Oregon State University. This work is entirely Unclassified and derived from open sources.

W. C. Thompson (✉)
Oregon State University Extension Service, Corvallis, OR, USA
e-mail: wiley.thompson@oregonstate.edu

F. Galgano (ed.), *The Environment-Conflict Nexus*, Advances in Military Geosciences, https://doi.org/10.1007/978-3-319-90975-2_7

As the entire region is very complex, a holistic understanding of the operational setting demands: a full appreciation of the ability for stakeholders to hold regional actors accountable; an examination of key major environmental issues; and an analysis of regional security risks through a modified approach for assessing non-land based environmental security. This chapter examines these issues, models outcomes if no intervention is offered, and recommends contexts where China can be influenced and may be more willing to amend their activities in the region.

Keywords China · Commercial fishing · Conflict · Coral · Coral reef · El Nino · Environmental scarcity · Environmental security · Fish · Indonesia · Island building · Malaysia · Paracel Islands · Philippines · Pratas atoll · Resources · Sea level · Security · South China Sea · Spratly Islands · Thailand · Transnational environmental issue · United States · Vietnam

1 Background

The interplay of resources, access, and security in the South China Sea (SCS) are multifaceted and offer a very interesting geographical problem. This is a complex region when viewed from any aspect. Topics such as political geography, ethnicity and language diversity, terrestrial and aquatic resources, climate change, historical conflict, and colonial legacies just begin to cover the variety of considerations which should be brought in for analysis. The significant contemporary environmental issues, the environment for advocacy or lack of, and a unique environmental security context, one that is largely aquatic, makes this region a compelling area for case study.

From a perspective of United States (U.S.) interests and security policy, military leaders and government policy makers have sought to understand what motivates or can be used as a lever against the People's Republic of China (PRC), specifically in the context of their expansion into the SCS along the Nine-Dash Line (Fig. 1)? Will a carrot or a stick approach, or a combination of both work best? Can pressure for environmental stewardship offer security benefits for other actors? Who best delivers this leverage and is it delivered directly or indirectly? Environmental change has played a significant role in internal stability and conflict within China over the past millennia when variations in climate results in cooling trends that negatively impacted agricultural and correlated with warfare frequency and dynastic change (Zhang et al. 2007). More modern times have witnessed socially constructed famine as in the case where Mao did just that from 1958 to 1962 to shape policy, practice, compliance, and output during his Great Leap Forward. Can environmental issues external to China still resonate with the domestic population and thus be used as levers?

Leaders of the PRC respond to pressure from environmental issues and have changed policy in the recent past. One interesting change the government of China has made, based on pressure from environmental groups, was to stop serving

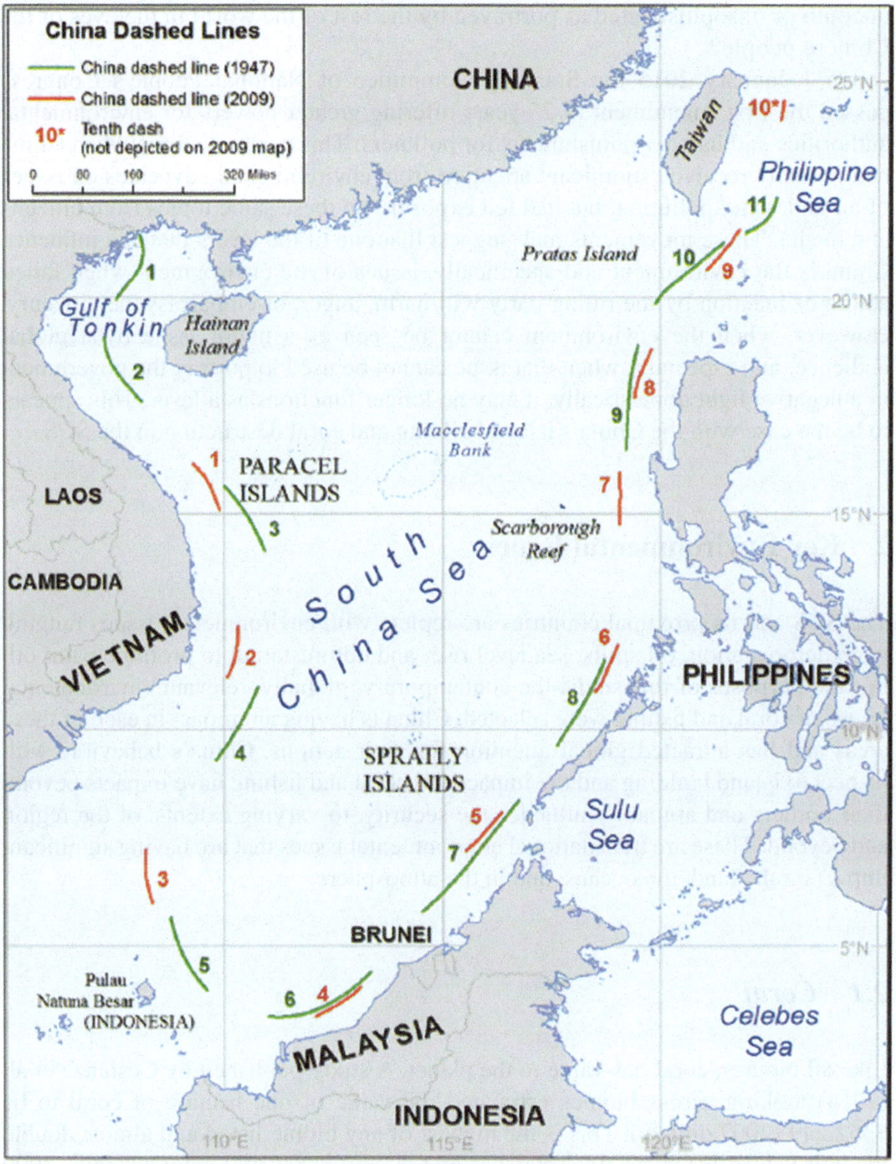

Fig. 1 Map depicting features of interest in the South China Sea to include China's 1947, 2009 and 10th dashed lines. (Source: U.S. DOS (2014))

shark fin soup at state dinners (Eriksson and Clarke 2015). Most impressive is that this resulted in a decrease in the consumption of shark fin soup by a growing middle class at a time when they could afford it. Was it concern for the environment or could it have been the desire for the government not to appear backwards, uncouth or unsophisticated as portrayed by the rest of the world in the eyes of the Chinese people?

On 1 January 2015 the Standing Committee of National People's Congress passed the first amendment in 25 years offering greater powers for environmental authorities and harsher punishments for polluters. This came at a time when China had not only received significant attention from environmental advocates on issues of air and water pollution, but had felt exposure on these same topics from citizens and media. These movements may suggest that one of the levers that can influence China is the environment and specifically, issues of the environment when failed action or inaction by the ruling party will harm, anger, or embarrass the citizenry. However, when the environment cannot be seen as a major issue by a global audience, and especially when that issue cannot be used to portray the government in a negative light domestically, it may no longer functions as a lever. This appears to be the case with the China's island building and coral destruction in the SCS.

2 Key Environmental Issues

The SCS and its proximal countries are replete with environmental issues ranging from deforestation, oil spills, sea level rise, and deforestation to promote palm oil. For the purposes of this study, the contemporary, globally–relevant environmental issues of coral and fishing were selected. China is having an impact in each of these areas and has attracted global attention for their actions. China's behaviors with respect to island building and the impact and coral and fishing have impacts beyond their borders and arguably influence the security, to varying extents, of the region and beyond. These are transnational environmental issues that are having significant impact on the land, the oceans, and in the atmosphere.

2.1 Coral

Like all biomes, coral has value to the planet. A study published by Costanza et al. (2014) looking across biomes estimated the value of one hectare of coral to be $352,249 (2007 dollars). This is the highest of any biome listed and almost double the unit valued listed for tidal marshes and mangroves. China now occupies 3000 acres (1214 hectares) of artificially constructed island space and has built land at a pace that is 17 times greater (over the last 20 months) than all other claimants in combined efforts over the past 40 years (U.S. DoD 2015). Damage to coral from destructive fishing practices is not new to the region and not unique to this area. Yet, beyond calls by the U.S. and to an extent the government of the Philippines, there has been little demand for accountability and for a halt to these destructive practices.

One of the few calls for accountability from the academic community is Dr. John McManus of the University of Miami's Rosenstiel School. His work has been the most cited in efforts to highlight the damage being done to coral reefs in the SCS. McManus is the principle investigator for the Peace through Conservation in the SCS project, a project that dates to the 1990s (McManus 1993). He has expressed concerns for an already overfished–fish region and the impact on island expansion on Scarborough Atoll, Pratas Atoll, the Paracel Islands, and the Spratly Islands corals to a point beyond which they will be able to recover. Additionally, he cites concerns over potential for armed conflicts in the past, arrests of fishing crews, and "*the constant threat of escalation to violence stemming primarily from competition over dwindling fishery resources*" that the surrounding coastal populations live under (McManus 2016b p 1).

McManus teamed with other colleagues to examine the human impacts on reef development for geopolitical and military purposes (Mora et al. 2016). They noted the unique value of isolated and uninhabited SCS islands and atolls calling them "reefs of hope," protected from invasive species, agricultural runoff, and pollution from sewage (Mora et al. 2016 p 1). Their emphasis on the value of isolation, which these reefs offer, are a compelling argument for coral preservation as a unique marine environment in the SCS. Supporting this suggestion Alan Freidlander, a biologist at the University of Hawaii, was quoted in an interview (Ranada 2015) noting that "*dredging and building on coral reefs in the SCS is causing irreparable damage to one of the most diverse ecosystems on earth*."

When it comes to coral destruction, China has an interesting ability to self-advocate. Beijing actually suggests that their island building efforts are a Green Project (Allen-Ebrahimian 2016). Although the construction on reefs and the harm it causes is well-documented, Beijing claims no harm is being done. This is an interesting narrative, in which the Chinese claim that their island building techniques simulate the natural processes of weather as "*sea storms blowing away and moving biological scraps which gradually evolve into oasis on the sea*" (Allen-Ebrahimian 2016 p 1).

As a result of climate change, Australia's Great Barrier Reef Marine Park Authority specifically noted that coral reefs are currently at increased risk of damage by rising sea temperatures, ocean acidification, and extreme weather events (AGBRMPA 2016). Changes in the ocean environment can make growth and recovery from damage by humans even more difficult. Ocean acidification from increased carbon dioxide (CO_2) harms corals as it slows the rate of calcification (i.e., skeleton formation) which forms the hard corals. The skeletons that do form under these conditions are also weaker. Work by Chen et al. (2015) examined the impact of increased sea surface temperature range and increased atmospheric carbon dioxide on coral as a result of climate change. Their findings confirm commonly accepted thought that both increased temperature and CO_2 result in a decrease in coral coverage and further suggest that economic losses in the global coral reef value as driven by climate change may range from U.S. $3.95 to U.S. $23.78 billion annually (Chen et al. 2015).

The recent 2015–2016 El Niño, also known as the third global bleaching event, has resulted in coral bleaching on a global scale. The El Niños of 1998 and 2010 had damaging effects on global coral populations, with 16% of reefs being killed world–

wide during the 1998 event (Underwater Earth 2016). Data suggest that the 2015–2016 event is on track to impact 38% of the world's coral reefs resulting in 12,000 square kilometers (4633 square miles) being killed (XL Caitlin 2016). Research from NOAA suggests that the bleaching event could affect the livelihoods of 500 million people, putting at risk income worth over USD $30 billion and also damaging coral reefs which buffer coastal communities from storms, a service that is critical, yet difficult to monetize (NOAA 2016).

The government of the Philippines remains the only major regional actor to speak out against China's destructive land reclamation practices and its commercial fishing fleet's illegal and destructive activities. In 2015, the Philippine Department of Foreign Affairs (DFA) released a statement calling out China's reclamation activities as causing "*irreversible and widespread damage to the biodiversity and ecological balance of the [SCS]/West Philippine* Sea (WPS)" (ROP 2016 p 1). The DFA pointed out that China's activities have so far caused destruction to over 300 hectares of coral reef systems, amounting to an annual economic loss of USD $100 million as well as constituting a threat to the livelihood of peoples and communities in the littoral countries. The statement also criticized China for being tolerant of violations under the Convention on Biological Diversity (CBD) and the Convention on International Trade in Endangered Species of Wild Fauna and Flora (CITES) as their fishermen poach giant clams, green sea turtles, and other endangered species. While raising critical environmental issues, the statement noted the security issues like China's pursuit of the Nine-Dash Line (Fig. 1) through reef destruction and island building are of significant regional importance (ROP 2016). This affirms that while environmental issues are present, security issues may be at the root of the matter for many.

2.2 Fish

In 2012, the SCS accounted for approximately 12% of the global seafood catch (Seas Around Us 2016). This is a significant number, given the size of this body of water. Over–fishing and destructive environmental practices are poised to negatively impact the value of fishing stock. If nothing changes, a report from the University of British Columbia suggest that the SCS could lose nearly 60% of its stocks by 2045 (Sumaila and Cheung 2015). China is a major stakeholder in the regional fishing economy as they account for about 45% of the landed value of fish taken out of the SCS (Sumaila and Cheung 2015). Overall China has seen its rate of annual marine capture more than quadruple from 7% in 1980 to 32% in 2013 (Baker et al. 2016).

Of particular concern has been the practices of Chinese commercial fishermen (and others) who use destructive techniques to include: dynamite fishing, cyanide poisoning, and bottom trawling. Giant clams are particularly valuable as they may fetch upwards of USD $1000–2000 each. The Chinese fishermen, mostly from Tanmen, harvest these clams which have increased in value as pressure in the illegal ivory trade has made them a sought–after substitute for ivory carvings (McManus 2016a). The SCS is also home to green sea turtles, which are endangered and hawksbill turtles, a critically endangered species. Further, much of the world's tuna stock are born in the SCS, making this a critical source region for this valuable commodity.

The impacts of Chinese fishing, coral destruction, and island building can have global, regional, and local impacts. This has already been recognized by fishermen and scientists. The reclamation of 769 acres (of Philippine interest) and the impact on fish populations could cost the Philippine economy and its fishing industry USD $110 million annually, according to the Bureau of Fisheries and Aquatic Resources of the Philippines (Ranada 2015). In Ranada's article she cities, Filipino scientist Dr. Edgardo Gomez, as he discussed the ecological significance of the West Philippine Sea noting a 2011 study which found that coral reefs in those waters were critical to the fish populations in the neighboring Sulu and Sulawesi Seas. Further, the SCS coral reefs provide spawning grounds and nurseries for a variety of marine species, demonstrating their value to ecosystems and economies outside their waters.

As Chinese *per capita* consumption of fish is now more than double that of the rest of the world, there may seem to be a compelling narrative available to offer in defense of Beijing's expansionist activities. However off shore fishing is only one method of providing fish for trade or domestic consumption. In a 2014 report, the UNFAO noted that "*China has been responsible for most of the growth in fish availability, owing to the dramatic expansion in its fish production, particularly from aquaculture*" (UNFAO 2014 p 3). This weakens the argument for a need to expand fishing operations into the SCS. Will the Chinese see fish as a strategic commodity to be defended as suggested by Dupont and Baker (2014)? The actions to militarize their commercial fleet suggest that they do (Rajagopalan 2016).

3 An Environmental Security Assessment

In the greater context, events in the SCS can be viewed through the lens of environmental security. In a recent publication focused on environmental security and migration in Africa, the United Nations Environmental Programme framed environmental security as "*the process of establishing the security of those environmental factors – water, soil, air, vegetation, biodiversity, climate and others – that are prime components of a nation's environmental foundations that ultimately underpin all its socioeconomic activities, and hence its political stability*" (UNEP 2017 p 1). Others suggest that environmental security can be framed in the context of a statist viewpoint or alternatively as environmental change that begets social change, which in–turn creates conditions for conflict, even down to the community level (Dabelko and Dableko 1995). A more recent definition of environmental security refers to a broad range of security issues intensified by environmental factors and suggests that environmental stress has the potential to trigger violent

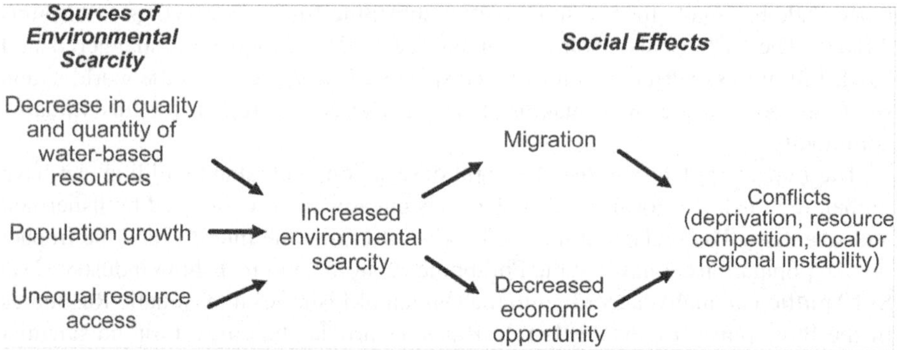

Fig. 2 Path from environmental scarcity to deprivation conflicts. Current conditions in the South China Sea are more likely to follow the lower path. Climate change can exacerbate conditions being seen today. (After Homer-Dixon (1994))

conflict (Galgano and Krakowka 2011). Homer–Dixon, a pioneering scholar and researcher in this area prefers not to use the term security, but instead focuses on links between environmental stress and violence, noting that violence is easier to identify and measure while acknowledging that security has been used to inflate the import of issues in an attempt to make them more competitive for "*public and policy maker attention*" (Homer-Dixon 1995 p 189).

While definitions or the application of terms may vary, the focus of a change in the environment and competition for resources and their correlation to violence remains constant throughout the literature. What is germane to this case study and is strongly emphasized in works by Dabelko and Dableko (1995) and Homer-Dixon (1995) is that any analysis must also incorporate the motivations of other actors, climate change, and a range of social, political and economic considerations. The setting of the SCS as a case study for environmental security is unique. Previous studies have focused on land-based events in Rwanda, Sudan, El Salvador, Haiti, Peru, the Philippines, and South Africa. Yet existing models should apply in a rather straightforward manner within the context of what is occurring, regardless of the terrestrial or aqueous environment, with models incorporating climate change implications being even more applicable.

One of the frameworks that can be used to model actions and outcomes in the context of environmental security (scarcity) was developed by Homer-Dixon (1994) and used in his work entitled Environmental Scarcities and Violent Conflict: Evidence from Cases. An adaptation of the framework he developed is depicted in Fig. 2. By following specific pathways in his framework and modifying terminology, his approach can be adapted to model the scenario being played out in the SCS. As previously noted, he would substitute scarcity for security in his writings. The activities creating environmental scarcity and the manifested effects as seen in the SCS have a distinct geography. Though Homer-Dixon (1994) does not account specifically for climate change, subsequent models suggest that the impacts of climate change can exacerbate anthropogenically created scarcity and social effects.

Table 1 Matrix depicting stakeholder and corresponding interests

Stakeholder	Interests and needs
China	One of the top SCS fish production and export countries. Important sector in economy. Significant distant water fleet.
Hong Kong	Local waters largely unregulated and unmanaged. Important fish stocks over–exploited and depleted
Philippines	Much of the fish is consumed locally. Decline in municipal and coastal fisheries due to heavy fishing pressure
Indonesia	Fisheries in critical condition. Illegal fishing is extensive
Thailand	One of the world's top fish producing countries. Fishing is important for domestic consumption. Primary source of protein for the population, especially in remote villages
Malaysia	Fishing critical to food security. Important for income and employment in rural areas
Vietnam	Critical to economy, 90% of catch is taken from shallow, inshore waters. Increasingly focusing on off–shore tuna market

The impact of fish availability has a distinct geography and varies in the communities it affects. (Derived from source material in Sumaila and Cheung (2015))

Throughout the SCS basin evidence and observations suggest that all three of the sources of Homer-Dixon's (1994) environmental scarcity are evident: (1) a decrease in quality and quantity of renewable resources; (2) population growth; and (3) unequal resource access. There may be a decrease in the quality of resources but there certainly has been a decrease in the quantity of renewable resources in the basin. When properly managed, the oceans can provide diversity of renewable resources for local consumption and in support of export economies. Fish stocks are one of the most exploited of these resources. Noted that by 2045, 60% of the fish they studied in the SCS will generate less catch with the outcome being serious food security implications and 55% of the fish will generate less value resulting in significant economic consequences (Sumalia and Cheung 2105). The authors also indicated that there is concern with overfishing and habitat destruction and share other data which suggest significant declines in fish population by species and in biomass overall (Sumalia and Cheung 2015). Both will influence livelihoods of local fisherman from coastal communities in the Philippines, Indonesia, Malaysia, Vietnam, and China, and the economies and markets which are symbiotic with the fish harvest. As noted by Sumaila and Cheung (2015) each stakeholder has its own interests and needs. These interests and needs are listed in Table 1.

While the impact on export economies, like those of Thailand, Vietnam, and especially China is significant, it is the local fishing communities that may be most affected, especially when there are few alternative livelihood options. The pressure placed on these small, near–shore fishing enterprises is even greater as China develops its commercial fishing fleet, transforming it into an expeditionary force with military training and supporting it with fuel and ice subsidies (Rajagopalan 2016). This clearly suggests that the intent is to use this fleet far from China's coastal waters, putting greater stress on distant marine resources and making it very difficult to manage regional fishing stocks with existing laws and enforcement mechanisms.

The expeditionary nature of China's commercial fishing fleet, combined with their rejection of the 2016 U.N. Tribunal ruling, suggests that China sees the waters around the built islands, specifically the Spratly Islands and Scarborough Shoal, as exclusive economic zones and that they will continue to exploit resources in these areas (Fig. 1). These islands are distant from China, but very close to the Philippines and Malaysia. This fits Homer-Dixon's (1994) framework as large militarized commercial fleets fishing in waters traditionally used by local fisherman will create unequal resource access as another source of environmental scarcity. With other countries unable, or unwilling, to support their local fishing industry in the same manner as the Chinese, the potential for Chinese fishermen to outcompete local fishermen in their own waters exists. Further, the militarization of the Chinese commercial fleet could similarly be seen as a power move to coerce other fishing vessels and dominate the marine harvest. When viewed both from a spatial sense and one of power, the conditions are set for unequal resource access.

China, Vietnam, and Indonesia are seeing rapid population growth along their coastal zones with Vietnam's coastal population growth expanding 0.02% faster than the rest of the country (Hinrichsen 1999), with Vietnam overall adding half the absolute number of people per year as Indonesia, but doing so on land area that is six times smaller. The growth rate in the Philippines province of Palawan was 4.65% from 2000 to 2010, whereas the growth rate in Mindanao was less than half that at 2.14% (NEDA 2016). Currently there are many large, rapidly urbanizing populations living within the coastal zone. Coastal zones include the massive river deltas found throughout Asia, all of which are critical to local livelihood and home to many. Consider the staggering statistic that although coastal zones account for only 2% of land area, they contain 10% of the global population and 13% of the global urban population (McGranahan et al. 2007). Possibly of greater concern is that one–third of the world's population lives within 200 kilometers of the coastline and one–half lives within 260 kilometers of the coast (Hinrichsen 1999; World Bank 2010).

As populations in these areas increase, rising sea levels continue to place them, their resources, and livelihoods at greater risk from coastal flooding, storms, and salt water intrusion. Such predictions suggest that maintaining a sustainable fishing economy is all the more critical in coastal communities. In 2009, the World Wildlife Fund commissioned a study to examine climate vulnerability in 11 major Asian coastal cities. The reports shared concerns that future climate change and vulnerability will impact national and economic security as well as "*human health, food production, infrastructure, water availability and ecosystems*" (WWF 2009 p 3). In China, rivers, deltas, and the coastal zone have become polluted, forcing fishermen farther off–shore to find viable fishing grounds. This not only places them in competition with other Chinese fishermen, but also with fishing fleets from other countries.

The opportunities for environmental scarcity are more than just possible because conditions exist and are being seen by scientists and those who make a living from the sea. At this point Homer-Dixon's framework (Fig. 2), which was developed prior to the time that climate change was well integrated into the body of environmental research, points to conditions of scarcity. Evidence suggests that a warming climate and increased variability will negatively affect the quality and quantity of renewable resources in the SCS. Warmer waters will alter fish habitat and breeding grounds, possibly forcing migration. As previously noted, some fish populations are genetically connected to populations in other bodies of water, indicating that impacts in the SCS can have teleconnections, making resolution to this crisis not only even more compelling, but also suggesting that in the longer term, chronic environmental impacts may bring in other stakeholders from outside the immediate region. An increase in atmospheric CO_2 has resulted in an increase in carbonic acid in ocean waters, which is hostile to marine life, especially in the context of the regeneration of coral reefs. Changes in water chemistry can impact other marine species, resulting in possible decrease in population numbers or migration out of the basin, similar to the fish the migration of fish populations.

In Homer-Dixon's (1994) model (Fig. 2), increased environmental scarcity can lead to either migration/expulsion, decreased economic productivity or both. In this case study, decreased economic productivity has already been seen, but accurate data are not available. What could make this more problematic is the consideration that local fisherman, where the impact may be hardest felt, have significant investment in their fishing livelihood (i.e., equipment, market connections, home location), thus making supplemental livelihoods or conversion to alternate livelihoods all the more difficult. This is compounded by the lack of available land and the entry of land-based farmers into the aquaculture fishery economy. These constraints leave few options for those in the fishing fleet economy. As such, competition-induced scarcity, especially from a militarized and heavily subsidized foreign fishing fleet, could accelerate the scarcity and worsen the decreased economic productivity.

The question of expulsion or migration is more problematic. As noted above, there may be some economic disincentives to migration by fishing communities. However, this is an aspect that should be examined at the local level in the coastal regions of each country. Migration literature is replete with examples of the impact of distressed economic conditions pushing people from rural regions towards perceived opportunities in urban communities. Just as plentiful is the evidence that these opportunities are only perceived and the situation for these economic migrants does not improve. What might be readily evident in this current context is a denial of resource or opportunity on a very localized scale as militarized Chinese commercial fleets outcompete and even "bully" local fleets away from areas illegally claimed by China. Similarly, the presence of Chinese military boats and aircraft in the region where island building is taking place may result in the intimidation of local fishing fleets, with the result of expelling from their traditional fishing grounds. This could be seen as a localized, forced, economic migration.

Homer-Dixon's (1994) model suggests that social effects may include ethnic conflicts, *coups d'état*, or deprivation conflicts. Conflicts between regional navies, coast guards, and fishing fleets have already occurred. At this point there is little evidence that any state will weaken to the point that a violent overthrow of a government will occur because of environmental scarcity in the SCS. What could happen over time, however, as scarcities become more acute and the impacts on livelihoods are manifested across larger populations, is that disenfranchised communities may begin to lose confidence in their government's ability to maintain sovereignty over territorial seas and existing regimes may be replaced through the political process.

The more likely outcome, as suggested by Homer-Dixon (1994) is the triggering of a violent conflicts, which may manifest in the form of deprivation, resource competition, or as local or regional civil instability. In fact Homer-Dixon (1994) suggests that evidence demonstrates "*that environmental scarcity simultaneously increases economic deprivation and disrupts key social institutions, which in turn causes 'deprivation' conflicts such as civil strife.*" This fits the current scenario very well whereby a decrease in the availability of fish and intimidation by the militarized Chinese commercial fleet or Chinese military planes and boats will create conditions of economic deprivation in regional fisherman, especially those who are locally based. For example, while aquaculture in the Philippines does provide part of the fish production and would not be impacted directly by aggressive island building, it represents only 25.4% of the total catch (FAO 2016).

There is an uncomfortably constant stream of reporting on confrontation between fishing vessels and regional coast guards and navies. Reports, like the article posted by Stratfor (2016) entitled *Fish: The Overlooked Destabilizer in the SCS*, suggests this trend will continue as China sends its fleet away from its depleted shores towards the less depleted waters near the Philippines, Malaysia, and Indonesia. While all of these encounters have been resolved in a relatively peaceful manner to date, it may only be a matter of time until they are not. Further as economic deprivation builds, the impact on livelihoods and coastal communities will eventually manifest in negative outcomes as the social institutions are disrupted and stressed. This has longer-term implications and could, overtime if not addressed by government entities, have negative social, economic, political and eventually, security implications.

4 Conclusions

The value of using a model is in its ability to help us understand complex realities in a more simplified and applicable framework. By modelling and identifying critical attributes and paths we can also look for opportunities to intervene and shape or prevent actions that would lead to undesirable outcomes.

First what should be considered is what can be done to influence a country of 1.38 billion people that has a U.S. $7.0 trillion global economy that sits on a 3.7 million square mile land mass? The answer, at least partially, may lie in looking for what motivates China, and what it values. As noted by Homer–Dixon, assessing the environment and security must incorporate the motivations of other actors in the region in addition to change and scarcity in the environment. According to Dr. Alex Vuving, a faculty member at the Asia–Pacific Center for Security Studies, the interests in the region of most actors largely center on power, resources, and sovereignty, though they vary from country to country. When it comes to China, Vuving suggests that they are playing Weiqi or Go, not chess. Not seeking victory through checkmate, China is seeking victory through encirclement, territorial gain, and control. He further notes that China does not fear singular attacks as they have a history of carrot and stick diplomacy, selective punishment methods, and an "attack us once we will attack you 1,000 times" mentality. Evidence of this was displayed in the pre-emptive rejection of 2016 The Hague Tribunal findings ahead of the official release and then their resolute rejection following the official release.

When it comes to the motivations of regional actors Vuving suggests that China's objective is first and foremost power and that they are pursuing it in a Mahanian control of the seas in a total hegemonic fashion. The Philippines, he suggests, prioritize resources (i.e., oil and gas) first then, sovereignty. Malaysia values resources (i.e., oil and gas), then sovereignty while Vietnam is mostly focused on sovereignty. Brunei has the least issues of anyone, they would prioritize sovereignty above all. While it appears that there will be little chance of direct conflict between any state actor and China, there are already confrontations between navies and fishing vessels and the conditions are set for the risk of conflict, taking on a variety of manifestations between sub-state actors.

In his chapter entitled, *Geopolitics and the dragon's advance: An exploration of the strategy and reality of China's growing economic and military power and its effect upon Taiwan*, Dr. Clifton Pannell (2011), one of the foremost U.S. experts in China's human geography advised that from the Chinese perspective, "*victory will be based on the full support of a prosperous and contented population while engaging in strategic diplomacy in step with military preparations*" (Pannell 2011 p 361). All of these are clearly evident in the context of environmental issues internally and in the SCS. When the air becomes so polluted that the harm it produces is undeniable to Chinese citizens and global media attention makes emissions an embarrassing issue to the Chinese government, the government takes action.

Though less visible, the government has taken action to clean up Chinese rivers and coastal waters, while in the interim, providing fuel and ice subsidies to off-shore fishing fleets, giving them not only an alternative fishing opportunity and making them an implement of the national security apparatus in the process. Again the prosperous and contented population takes primacy and gets action. As Beijing has thumbed its nose at the Tribunal's ruling, the challenge, if the U.S. wants to make the environment an issue with any leverage, is how to get the global community to rally behind an environmental call to arms over rock and coral that amount to an almost immeasurable area of the SCS. With no check on China's behavior, either from a legal, diplomatic, or influence of domestic support/approval mechanism, consideration must be given to Dr. Pannell's argument that the Second Island Chain is soon to follow as military and technological capabilities and resources mature.

Given little apparent desire to challenge destructive policies from the environmental community and China's refusal to accept the Tribunal's ruling, policy makers may consider developing strategies from Homer-Dixon's framework to intercede and mitigate the risk of deprivation conflicts? Any strategy developed would be best applied in the early stages where *sources of environmental scarcity* are created: a *decrease in quality and quantity of renewable resources*; *population growth*; and *unequal resource access*. This will impact all scarcity generators except for population growth, which may be negligible from the perspective of populating the built islands.

The two sources of environmental scarcity that can be influenced through are a *decrease in quality and quantity of renewable resources* and *unequal resource access*. In order to mitigate the risk of conflict, economic opportunity and productivity must be secured. To do this, marine resources must be sustainably managed and reasonable and equitable access must be offered. U.N. Convention on the Law of the Sea (UNCLOS), while the source of contention with island building, may be a platform that can be used to find cooperation when it comes to managing the marine harvest. This could be further strengthened with the enforcement of UN Fish Stocks Agreements. Another mechanism that would assist in this effort is more concerted efforts to decrease illegal, unreported and unregulated (IUU) fishing. All of these suggestions would require acceptance, compliance and enforcement of these international agreements. However, unlike the coral destruction and island building, failed compliance is likely to get the attention and advocacy from environmental groups, something that has not happened with coral, but would have the same outcome. In this situation, sustainable fishing is primarily an environmental issue, with negative impact on livelihoods and species depletion being the outcomes of failed action.

Another opportunity to protect marine resources and create the conditions for equal access may actually come through denying or limiting access. Creating marine protected areas (MPAs) or "marine peace parks" as suggested over two decades ago by Dr. McManus could function as a mechanism to control and balance access to space. Environmental groups would be very unlikely to engage on issues related to UNCLOS disputes over territorial waters or exclusive economic zones which are how the island building disputes are seen right now. Sanctuaries or protected parks can function to limit access to fishing fleets and also limit China's island building activities. If the parks included islands already developed, given the Tribunal's ruling that China has no claim to these areas, the Chinese may be compelled, through "shamefare" by environmental groups to abandon these posts. Refusing to abandon the destructive activities associated with military occupancy (waste management, fuel/oil/chemical spills, and noise hazards), especially if they are now taking place in the middle of a marine sanctuary, may finally get environmental advocacy groups engaged.

In the end, the only way to make security issues in the South China into environmental issues is to truly make them environmental issues. If the U.S. and other interested actors can construct a framework of compelling environmental issues that leave no room for China's expansionist aims and makes the issues resonate with environmental primacy, then environmental advocacy may become a reality, as long as security remains masked as a distant and collateral outcome. Finally, analyzing security interests that intersect with the environment through validated frameworks can offer new perspectives, alternative hypotheses, and a more constructive voice in which to communicate with non-security minded audiences. Military strategists and policy makers should consider adapting these frameworks, when appropriate to develop a more holistic understanding of the complex issues they face. As the global population continues to increase and consumption rises along with it and as climate change modifies the environment, impacting opportunities for livelihoods, all stakeholders will eventually come to the realization that security issues and environmental issues are truly co-equals and those who viewed them in that manner all along were well ahead of the rest.

Literature Cited

Allen-Ebrahimian B (2016) Beijing Calls South China Sea Island Reclamation a 'Green Project'. http://foreignpolicy.com/2016/05/26/china-calls-south-china-sea-island-reclamation-a-green-project-spratly-islands/. Accessed 29 Dec 2017

Australia's Great Barrier Reef Marine Park Authority (2016) Managing the reef. http://www.gbrmpa.gov.au/managing-the-reef/threats-to-the-reef/climate-change/what-does-this-mean-for-habitats/coral-reefs. Accessed 29 Dec 2017

Baker R, Harnagel A, Rees E (2016) Fish, the overlooked destabilizer in The South China Sea. https://www.stratfor.com/analysis/fish-overlooked-destabilizer-south-china-sea. Accessed 29 Dec 2017

Chen P, Chen C, Chu L, McCarl B (2015) Evaluating the economic damage of climate change on global coral reefs. Glob Environ Chang 30:12–20

Costanza R, de Groot R, Sutton P, van der Ploeg S, Anderson S, Kubiszewski I, Farber S, Turner R (2014) Changes in the global value of ecosystems services. Glob Environ Chang 26:152–158

Dabelko G, Dableko D (1995) Environmental security: issues of conflict and redefinition environmental. Change Secur Program 1:3–13

Dupont A, Baker C (2014) East Asia's Maritime Disputes: Fishing in Troubled Waters. The Washington Quarterly 37(1):79–98

Eriksson H, Clarke S (2015) Chinese market responses to overexploitation of sharks and sea cucumbers. Biol Conserv 184:163–173

Food and Agriculture Organization of the United Nations (2016) Fishery and aquaculture country profiles: The Republic of the Philippines. http://www.fao.org/fishery/facp/PHL/en. Accessed 29 Dec 2017

Galgano F, Krakowka A (2011) The environment's influence on regional stability and conflict. PA Geogr 48(2):3–32

Hinrichsen D (1999) The Coastal population explosion. In The next 25 years: global issues. http://oceanservice.noaa.gov/websites/retiredsites/natdia_pdf/3hinrichsen.pdf. Accessed 29 Dec 2017

Homer-Dixon T (1994) Environmental scarcities and violent conflict: evidence from cases. Int Secur 19(1):5–40

Homer-Dixon T (1995) Environment and security. Correspond Int Secur 20(3):189–198

McGranahan G, Balk D, Anderson B (2007) The rising tide: assessing the risks of climate change and human settlement in low elevation coastal zones. Environ Urban 19(1):17–37

McManus J (1993) The Spratly Islands: a Marine Park? AMBIO J Hum Environ 23(3):181–186

McManus J (2016a) Chinese fishers destroy reefs as anti-ivory action boosts clam market. http://thecoraltriangle.com/stories/chinese-fishers-destroy-reefs-as-antiivory-action-boosts-clam-market. Accessed 29 Dec 2017

McManus J (2016b) Peace through conservation in the South China Sea. http://www.rsmas.miami.edu/groups/peace-through-conservation-in-the-south-china-sea/. Accessed 29 Dec 2017

Mora C, Caldwell I, Birkeland C, McManus J (2016) Past island building reviewed: dredging in the Spratly Islands: gaining land but losing reefs. PLoS Biol 14(3):e1002422
National Economic and Development Authority (2016) Palawan. http://mimaropa.neda.gov.ph/palawan/. Accessed 29 Dec 2017
National Oceanic and Atmospheric Administration (2016) El Nino prolongs longest global coral bleaching event. US Department of Commerce, Washington, DC
Pannell C (2011) Geopolitics and the dragon's advance: an exploration of the strategy and reality of China's growing economic and military power and its effect upon Taiwan. In: Galgano F, Palka E (eds) Modern military geography. Routledge, New York
Rajagopalan M (2016) China trains 'fishing militia' to sail into disputed waters. http://www.reuters.com/article/us-southchinasea-china-fishingboats-idUSKCN0XS0RS. Accessed 29 Dec 2017
Ranada P (2015) China reclamation poses P4.8B in economic loss for PH. http://www.rappler.com/nation/90878-china-west-philippine-sea-reclamation-fisheries-food-security. Accessed 29 Dec 2017
Republic of the Philippines, Department of Foreign Affairs (2016) Statement on China's reclamation activities and their impact on the regions' Marine Environment. http://dfa.gov.ph/index.php/newsroom/dfa-releases/5913-statement-on-china-s-reclamation-activities-and-their-impact-on-the-region-s-marine-environment. Accessed 27 Dec 2017
Seas Around Us (2016) Tools and data. http://www.seaaroundus.org/. Accessed 29 Dec 2017
Stratfor (2016) Fish: the overlooked destabilizer in the South China Sea. https://www.stratfor.com/analysis/fish-overlooked-destabilizer-south-china-sea. Accessed 29 Dec 2017
Sumaila U, Cheung W (2015) Boom or Bust: the future of fish in the South China Sea. http://www.oceanrecov.org/news/ocean-recovery-alliance-news/boom-or-bust-the-future-of-fish-in-the-south-china-sea.html. Accessed 29 Dec 2017
UN Environmental Programme (2017) Environmental security – African Ministerial Conference on the Environment. https://www.unenvironment.org/resources/report/amcen16egm5-environmental-security. Accessed 29 Dec 2017
UN Food and Agriculture Organization (2014) The state of world fisheries and aquaculture 2014. http://www.fao.org/3/a-i3720e.pdf. Accessed 29 Dec 2017
Underwater Earth (2016) The 3rd global coral bleaching event – 2014/2016. http://www.globalcoralbleaching.org/. Accessed 27 Dec 2017
US Department of Defense (2015) Asia-Pacific Maritime security strategy. US Government Printing Office, Washington, DC
US Department of State (2014) Limits in the Seas No. 143 China: maritime claims in the South China Sea. US Government Printing Office, Washington, DC.
World Bank (2010) Climate risks and adaptation in Asian Coastal megacities: a synthesis report. http://siteresources.worldbank.org/EASTASIAPACIFICEXT/Resources/226300-1287600424406/coastal_megacities_fullreport.pdf. Accessed 29 Dec 2017
World Wildlife Fund (2009) Mega-stress for mega-cities: a climate vulnerability ranking of major coastal cities in Asia. http://assets.panda.org/downloads/mega_cities_report.pdf. Accessed 29 Dec 2017
XL Caitlin Seaview Survey (2016) Scientists confirm global coral bleaching event for 2015. http://catlinseaviewsurvey.com/news/08-10-2015/scientists-confirm-global-coral-bleaching-event-for-2015. Accessed 27 Sept 2017
Zhang D, Zhang J, Lee H, He Y (2007) Climate change and war frequency in Eastern China over the last millennium. Hum Ecol 35(4):403–414

East Africa in World War I: A Geographic Analysis

Andrew D. Lohman

Abstract Most geographic analyses conducted during and after the First World War focused exclusively on the influence of physical geography on battles in Europe. In recent years, however, a growing body of research has emerged on the campaign fought in East Africa by the European powers, which includes the importance of the human landscape and environmental security. The war that raged across East Africa was linked strongly to competition for vital resources among Europe's great powers and it illustrates the most problematic outcome of the environment–conflict nexus: i.e., interstate war. The scope of military geography has expanded and contemporary perspectives have advanced beyond describing the effects of the natural landscape on warfare. They include incisive analyses of the cultural landscape and how human geography shapes, and is shaped by conflict. This paper provides a military geographic perspective of the East African campaign, and analyzes how environmental factors and human geography dramatically influenced the course of this conflict. This paper will focus on salient aspects of physical and human geography that were decisive during the campaign. This analysis suggests that the region's natural and human landscape inevitably compelled German and British forces to involve hundreds of thousands of Africans as soldiers and laborers; and that they suffered severe causalities and depredations because of this war.

Keywords Africa · Askaris · Burundi · Climate · Conflict · Disease · East Africa · Environmental security · Environment–conflict Nexus · Germany · Great Britain · Guerilla · Imperialism · Malaria · Military geography · Mozambique · Nationalism · Rwanda · Tanzania · World War I · Zimbabwe

A. D. Lohman (✉)
Department of Geography and Environmental Engineering, United States Military Academy, West Point, NY, USA
e-mail: andrew.lohman@usma.edu

F. Galgano (ed.), *The Environment-Conflict Nexus*, Advances in Military Geosciences, https://doi.org/10.1007/978-3-319-90975-2_8

1 Introduction

The nexus of the environment and conflict—that is, environmental security—is thought to be limited to contemporary interpretations of armed conflict. Nevertheless, examples of violent conflict with an environmental component are quite common throughout history—fighting over resources is as old as war itself. The environmental security paradigm suggest that environmental scarcity, resources, and the adverse effects of climate change will influence three forms of warfare: (1) simple scarcity conflicts; (2) relative deprivation conflicts; and (3) interstate warfare (Homer–Dixon 1991). This case study examines the extension of the First World War into East Africa. Although World War I was propelled largely by a much larger geopolitical struggle, the war in East Africa did entail a significant environmental component. The origins of Europe's Great War are linked strongly to the seminal causes of nationalism and long–standing antagonism among the great powers; however, it was also fostered by imperialism and competition for the markets and resources in Asia, Africa, and the Middles East. Africa, in particular, was vital to the industrialization of Europe because of its vast natural resources, and the colonization of Africa by the European powers was related to their desire to secure these important natural resources (Keegan 1998). Hence, the British invasion of East Africa was linked to the larger geopolitical struggle with Germany, but it was also compelled by their desire to secure the abundant resources in the region (Meyer 2006). This was a form of limited interstate warfare, fought by proxy forces and is consistent with contemporary environmentally–related conflicts such as those witnessed in Somalia, Rwanda, Syria, and Darfur.

The First World War began more than a century ago, yet new research continues to emerge on its causes, course, and consequences. Although most analyses continue to focus (primarily) on the war in Europe and the Western Front, there is a large, and growing historical literature on the Great War in Africa; and in particular, the campaign fought in East Africa, which Anderson (2004) has so aptly referred to as the *Forgotten Front.* While typically referred to as a sideshow, the war in East Africa in particular, had a profound impact on Africa and the world today. Geopolitically, it is considered to be the final stage in the so–called *Scramble for Africa.* Although German strategy in East Africa was rather limited and focused on tying up Allied forces, European colonial powers viewed it as a key contest to secure further control over territory, resources, and populations in Africa (Strachan 2004). This campaign is perhaps the most illustrative example of classic geopolitics, and a "*period of increasing competition between the most powerful states*" for territories and resources (Flint 2006 p 17). Although far fewer casualties were incurred in this theater than in Europe, Strachan (2004) suggests that the effects of this war disrupted millions of lives, particularly for Africans—an estimated two million served as soldiers or laborers, along with countless bystanders who were inadvertently involved in the conflict as it raged across their homelands.

Literature on this campaign ranges from first–person accounts to detailed analyses of individual battles. Collectively, these works provide meticulous historical records of how the campaign was fought and the geographic factors that contributed to its outcome. Yet, contradictory conclusions have emerged from these analyses. Hoyt (1981) referred to Colonel Paul von Lettow–Vorbeck's (i.e., the German commander) performance as perhaps the greatest guerrilla campaign in history; and von Lettow–Vorbeck's own memoirs are considered a primer for waging guerrilla war (Ofchansky 1990). In contrast, however, Strachan (2004) considers it one of the greatest calamities of the Twentieth Century, equating its demographic impact on par with that of the Atlantic slave trade. Notwithstanding these opposing perspectives, there is strong consensus that the physical environment had a profound influence on the course of the campaign. Many publications highlight, to varying degrees, the significance of the natural environment, and its vital natural resources, on the conduct of military operations in East Africa. While not specifically stated in these publications, the concept of scale differentiates the effects of the natural environment on soldiers and civilians as well as on the operations and maneuver of tactical units.

A review of these discussions, however, indicates that they are broad and highly generalized descriptions, characterizing the weather and climate as 'hot', 'humid', 'tropical', 'dry', or 'dusty.' Furthermore, historical accounts highlight the devastating effects of malaria and the tsetse fly on people and transport animals; but similarly provide vague descriptions of where malaria and tsetse prone areas were located within the area of operations. Strachan (2004 p 12) notes, "*Both the climate, with its switch from dry to rainy seasons, and the insect life, with its impact on the health of livestock and humans, were strategically decisive*." Although there has been a great deal written about this forgotten front, most publications are written from a military history perspective, focusing on events rather than employing incisive geographical analyses. These publications do highlight the significance of terrain, resources, and climate in the East African campaign, but the natural and human environments are clearly not the central focus of these analyses. This chapter, in contrast, examines the war in East Africa from the vantage point of military geography and the environment–conflict nexus, and it attempts to provide a greater understanding of the strategically decisive role that the region's physical and human geography played in this campaign.

2 Military Geography

Terrain and climate have influenced military operations throughout history (Winters 1998), but it was not until World War I that military geography emerged as a distinct subfield. While geographers played a crucial role in the development of terrain intelligence and on planning staffs, Johnson's (1921) detailed study of the battlefields of the First World War marked the emergence of military geography as an important field of study (Palka 2004). Johnson's (1921) study, which focused on the effects of topography and underlying geological structure on First World War battles, shaped the early direction and scope of military geography. This work also examined the vital influence of cultural landscapes on warfare as he described the important influence of agricultural patterns, roads, canals, and railways on the course of campaigns because they added strategic value to particular places. Additionally, Johnson's (1921) analysis also considered the role of population centers, and how these affected the battle lines that emerged across the Western European landscape.

However, since Johnson's (1921) publication, military geographers have continued to expand the perception of how geography influences military operations. Galgano (1998), building upon Garver's (1975) development of the environmental matrix, provides a detailed framework that examines the role that the natural environment (i.e., location, weather and climate, surface materials, hydrology, vegetation, landforms and surface structure, resources) and cultural landscapes (i.e., population, cultural groups, cultural components and institutions, settlement and land use, economy, networks, and cultural capabilities) play in affecting the course of military operations. Similarly, Doyle and Bennett (2002) provide a conceptual model of terrain in military history which explains its influence and change across time as the *Symbolic Landscape* (ancient rites, cultural significance, religious symbolism) and shapes the *Landscape of Battle* (battlefield resources, topography and battle, and mobility and trafficability), resulting in the *Iconic Landscape* as revealed in the cultural, archaeological, and anthropological legacies of new landscapes. This chapter does not attempt to provide a complete analysis of the war in East Africa. Rather, it focuses on select physical and human geographic factors from these frameworks that are considered the most salient: i.e., climate, terrain, population settlements, roads and transportation networks, and endemic diseases. This analysis suggests that East Africa's physical and human landscape compelled German and British forces to involve hundreds of thousands of Africans who had little interest or say in this conflict.

3 The War in East Africa

After the start of the war in August 1914, German and British leaders in the African colonies attempted to maintain neutrality, as specified by the Berlin Act of 1885. Inevitably, however, military commanders saw it as their duty to engage the enemy regardless of the neutrality clause and the war spread over a vast area of Africa to include modern–day Tanzania, Rwanda, Burundi, Mozambique, Zimbabwe, and portions of southern Kenya (Strachan 2004; Anderson 2004). German East Africa, where the majority of the fighting took place, was an area of nearly 384,000 square miles (i.e., about as large as France and Germany combined), with approximately eight million Africans and 5000 European settlers (Sibley 1971). Fig. 1 illustrates the operational area and depicts the routes that German forces used during the campaign, which included a remarkable fighting withdrawal against vastly superior forces. In so doing, they tied up more than 150,000 British, Indian, South African, and British colonial African troops during the course of the war—remarkably, they did this with a force that never exceeded 15,000 soldiers (Abbott 2002). A rough

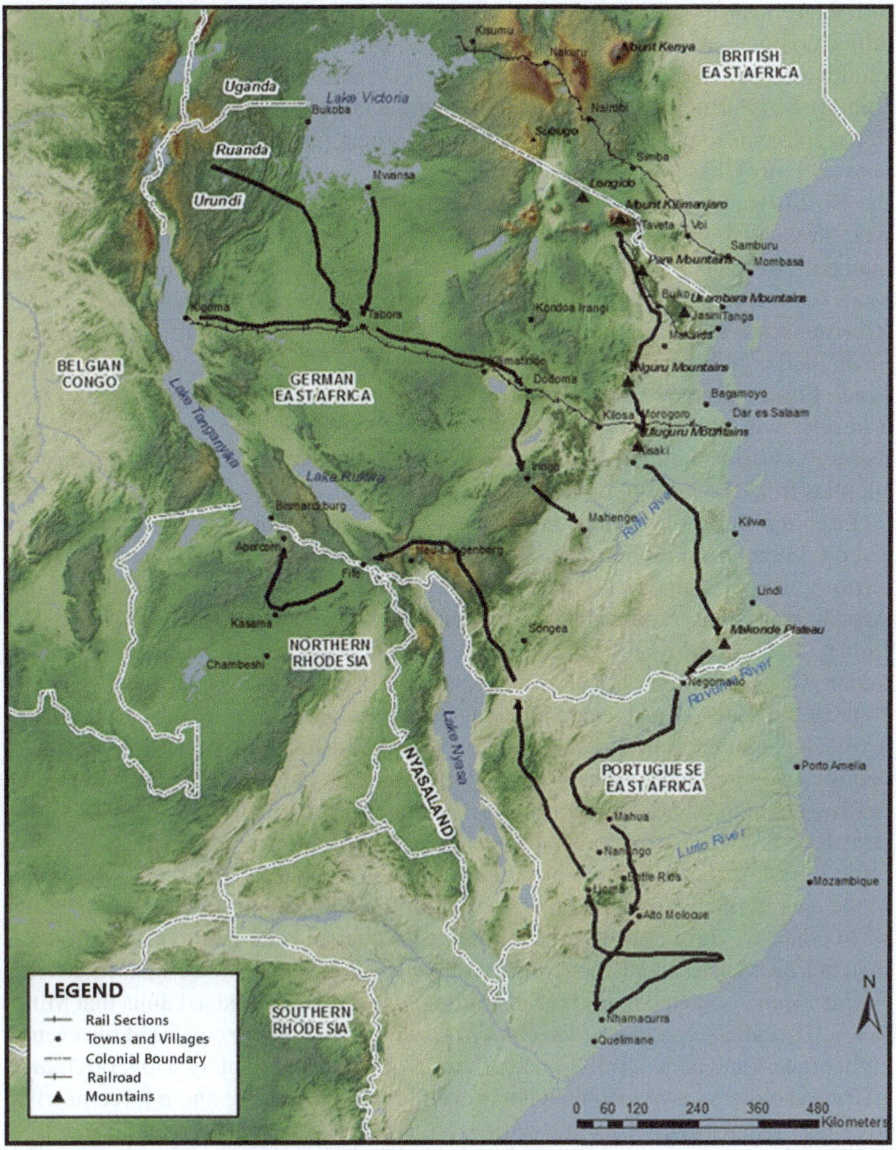

Fig. 1 The East Africa theater of operations in World War I. The black lines indicate the routes followed by the German forces during the course of the war, from Fall 1914 until November 25, 1918. Map by author after Sibley (1971) and Miller (1974)

estimate of the distances covered over the course of this campaign is approximately 4500 km during the four years of conflict.

While the German forces were able to take advantage of the geographic factors at the local or tactical scale, at the strategic scale they were at a significant and ultimately insurmountable disadvantage. German East Africa was surrounded by Allied colonies, with Britain in the north (i.e., British East Africa and Uganda) and southwest (i.e., Northern Rhodesia and Nyasaland), Belgium in the west (the Congo), and Portugal to the south. Africa's great lakes along the western boundaries of the colony also played a role in isolating German East Africa, as the British were able to control these waters early in the war. They were essentially surrounded with the British Navy in control of the Indian Ocean, so von Lettow-Vorbeck's forces were strategically isolated from Germany. A few supply ships carrying essential war materials were able to infiltrate the British blockade, but were insignificant to the total requirement needed by German forces to sustain a prolonged conflict (Christensen 2003). Additionally, by September 1916, the British also seized control of the major ports and towns along the coast (i.e., Tanga, Dar es Salaam, Kilwa, Lindi, Fig. 1), thereby further hindering German resupply efforts (Sibley 1971). As a result, von Lettow-Vorbeck's forces were compelled to supply and equip themselves from captured enemy stocks, as well as purchasing and commandeering supplies from the local population.

From the start of hostilities in August 1914, until 25 November 1918, when von Lettow-Vorbeck finally surrendered, the combatants traversed a large expanse of territory, which included a broad gradient of climate types and biomes. Thus, the African war is often described as the antithesis of the Western European theater (Sibley 1971; Keithly 2001). The Western Front was characterized by positional, trench warfare between large armies operating in a confined geographic space. In contrast, the East African campaign was a war of maneuver, fought by small forces across large territorial spaces (Sibley 1971). British forces in Africa tended to operate in larger formations (i.e., battalion and brigade–sized) to envelope and destroy German units (Anderson 2004). Though operating at this scale allowed the British to bring superior forces to bear, it also required a much greater logistical effort to sustain these forces, which became increasingly difficult because the effects of the operating environment made it difficult to maintain lines of communications.

In contrast, von Lettow-Vorbeck directed a guerrilla campaign, mounting raids then falling back to avoid encirclement or decisive engagement. German forces (the *Schtuztruppe*) operated in logistically autonomous, company–sized units that Miller (1974, 18) called a "*self-contained micro-army.*" *Schtuztruppe* consisted of seven or eight officers and noncommissioned offers (NCOs), approximately 150–200 *Askaris* (African soldiers), two machineguns, several hundred porters, and a force of tribal irregulars (*ruga-ruga*). Together with a surgeon and support personnel, these field companies were almost uninhibited by traditional lines–of–communications, which afforded them a tremendous degree of mobility (Miller 1974). This organizational structure, and the advantages it provided, was developed before the war in response to local uprisings against German rule. Unlike the large British units, German formations retained agility and were far less affected by factors of distance, poor transportation networks, and the imposing physical landscape.

Several other factors differentiate East Africa from the Western Front. Artillery, which played such a significant role, was "*the most common cause of casualties when in the trenches*" (Ellis 1976 p 61). Once out of the trenches, the machine gun emerged as the largest casualty–producing weapon (Ellis 1976). In East Africa, however, artillery was limited. For example, the largest guns available to support German forces were 4.1–in cannon removed from the sunken cruiser *Konigsberg*. Logistically, the widespread use of artillery was limited too by the insufficiency of transportation networks. The machinegun, however, was a decisive weapon in Africa.

The armies of the Western Front employed a great deal more terrain intelligence, which Doyle and Bennett (2002) describe as a database of information on the topography and landscape in the form of maps, plans, and photographs. They of course had a great deal of first–hand knowledge and experience in the region. Chasseaud (2002) indicates that Britain, France, and Germany collectively produced approximately 850 million maps to assist in operational planning and intelligence purposes. In contrast, British and German forces were significantly limited by the lack of African maps, and thus found it difficult to understand the region's terrain (Miller 1974; Sibley 1971). British and German memoirs (e.g., Smuts 1919; Christensen 2003; von Lettow-Vorbeck 1990; Meinertzhagen 1960) describe the appalling lack of maps and the general inability of commanders to visualize the terrain. Maps that did exist were generally inaccurate and outdated, and at too small of a scale to be effectively useful for planning tactical operations (Strachan 2004; Sibley 1971). As von Lettow-Vorbeck (1990, 56–7) noted in his memoirs of the campaign, "*It is, of course, always difficult to make plans for an action in country so totally unknown to one without the aid of a map*." Notwithstanding the shortage of good maps, the Germans possessed much better knowledge of the terrain than did their British opponents.

Of particular importance, von Lettow-Vorbeck benefitted from a better personal sense of the terrain. Upon taking command of German colonial forces in January 1914, he immediately began a personal tour of the region. Traveling through most of German East Africa by rail, road, bicycle, and foot, he gained an innate sense of the region's terrain. Furthermore, his German officers and NCOs had collectively spent several years in the colony and together with their locally–raised *Askaris*; they had a terrific understanding of the nature of the terrain and its inherent advantages and disadvantages. British forces, in contrast, were initially composed of only three locally–raised battalions of the King's African Rifles (KAR), and the majority of Commonwealth forces through early 1917 were composed of British, Indian, and South African units with little experience in the region. These units, and their leaders, lacked first-hand knowledge of the terrain over which they would fight (Meinertzhagen 1960). While these factors clearly differentiated the war in East Africa from that which was fought on the Western Front, it was the war of maneuver across the vast expanse of territorial space—a space with highly differentiated terrain and climate gradients—that truly distinguished this theater from the trenches of Western Europe.

4 War of Maneuver

As the fighting began in East Africa, von Lettow–Vorbeck sought to defend the colony by stationing *Schutztruppen* along the borders; however, he also undertook offensive actions to gain the initiative. The region of most immediate concern was the colony's northern boundary, centered on the rail line running between Tanga and Moshi at the base of Mount Kilimanjaro (Fig. 1). This was the most agriculturally productive and heavily populated region of the German colony (Miller 1974). Furthermore, its proximity to British East Africa and the vital Mombasa–Nairobi railroad provided the German command an opportunity to put British forces on the strategic defensive by directing attacks along this rail line.

von Lettow-Vorbeck's first significant operation was the seizure of Taveta, in British territory, situated in the strategic gap between Mount Kilimanjaro (19,341 ft) and the Pare Mountains (≈ 6000 ft). This gap was a critical corridor between British and German East Africa and was the key avenue through which British forces would have to pass as part of any attempt to attack into German East Africa (Miller 1974). Once in control of Taveta, German forces operating in small units conducted raids against the British rail line and ambushed British patrols. The British, hard pressed to stem these raids, dispersed their troops in small posts and forts along the railroad, but lacked sufficient forces to secure it and maintain a force large enough for offensive operations (Meinertzhagen 1960). Consequently, the British command requested assistance from the Colonial Office and two expeditionary forces were dispatched to East Africa from India (Anderson 2004). One force, the Indian Expeditionary Force C (≈ 2000 soldiers) arrived in Mombasa in early September 1914, and was dispatched to the Kilimanjaro area to secure the rail line. The second, Indian Expeditionary Force B (≈ 8000 soldiers), under the command of Major General Aitkin, was directed to conduct an assault landing at Tanga in early November 1914. The objective of this operation was to secure that vital port and proceed northwest along the railroad to catch von Lettow-Vorbeck's forces in a pincer. The German commander, however, through advanced warning, was able to concentrate several companies in Tanga to counter the landing. Aided by tactical errors and poor leadership, as well as a lack of Allied terrain intelligence, von Lettow-Vorbeck won a resounding victory. Furthermore, he forced the British to evacuate after 3 days of fighting, leaving behind a large supply of weapons, ammunition, and other supplies.

Following this success, von Lettow-Vorbeck's forces attacked north along the coast, capturing some British territory, including the town of Jasin, but it was a hard-won victory. After considerable expenditure of materials and supplies, including precious ammunition and the loss of a significant number of German officers and NCOs, von Lettow-Vorbeck decided that a guerilla campaign was the best way to conduct operations in East Africa (von Lettow-Vorbeck 1990; Hoyt 1981). Throughout 1915, his forces continued to raid the British railroad. However, a subsequent build–up of British forces, comprised mostly of South African and Indian forces, and the assumption of command by offensive–minded South African General Jan Smuts began to turn the tide against the Germans.

von Lettow-Vorbeck settled on a strategy that avoided direct battle and decisive engagements, and instead he relied on small unit, hit–and–run raids against the British. Similarly, General Smuts sought to avoid large–scale and costly direct attacks, and opted instead for a war of envelopment. In so doing, he hoped to maneuver around German forces, and encircle or annihilate them (Miller 1974; Farwell 1986). For both commanders, their strategies were predicated on preserving their forces and limiting casualties, but for very different reasons. Economy of force was essential to von Lettow-Vorbeck because he was isolated from reinforcement and resupply, and there were no chances to replace German officers and NCOs lost through combat. Smuts too, sought to prevent high casualties among his troops, particularity his South African forces. However, he was motivated by political as well as military factors. His South Africans could potentially be quickly replaced, but Smuts had political aspirations and did not want to be labeled as a butcher upon his return to South Africa following the war (Meinertzhagen 1960).

Smuts only served as the British commander from February 1916 to January 1917, however his replacement General Jacob van Deventer (also South African) continued Smut's strategy to maneuver the Germans into a position in which they would have to fight a decisive battle or capitulate. Unfortunately for the British, von Lettow-Vorbeck would not allow himself to be maneuvered into such a position. Although he faced Allied forces on several fronts, and was severely outnumbered, he enjoyed the advantages of interior lines. More importantly, he was operating in an area within which he and his officers were thoroughly familiar. These advantages were compounded by the flexible organizational structure of the *Schutztruppen* and their ability to live off the land. These important factors enabled him to choose the time and place of battle and retain a significant element of initiative despite being on the strategic defensive (Sibley 1971; Miller 1974).

The German defensive strategy was further aided because Allied forces failed to effectively coordinate their offensives and were impeded by a lack of effective communications. Furthermore, significant antagonisms existed between British, Belgian, and Portuguese commands (Anderson 2004; Cann 2001). Nevertheless, by March 1916, Smuts' flanking maneuvers in the north had forced von Lettow-Vorbeck to withdraw south from the Taveta gap to avoid encirclement. This began the German fighting withdrawal through German East Africa, which would eventually take von Lettow-Vorbeck and his remaining forces into Portuguese East Africa, as far south as Nhamacurra, before reentering German East Africa and proceeding into Northern Rhodesia, where he finally surrendered two weeks after the armistice in Europe.

Clearly, the scale of the theater of operations gave von Lettow-Vorbeck the operational space to maneuver, avoiding decisive engagements, and enabled him to avoid being confined to one place in a decisive battle. At the height of its strength, the German forces numbered approximately 3,000 Europeans and about 12,000 *Askaris*, whereas British forces had over 27,000 troops, with another 2,000 in Uganda and 2,500 in Northern Rhodesia. Additionally, the Belgians fielded over 12,000 troops along its colonial borders (Abbott 2002). Portuguese attempts to secure German territory in the south were easily repulsed because of logistical and leadership failures (Cann 2001). Of course, these numbers are only snapshots at particular times; and over the course of the entire conflict, more than 150,000 British, South African, Indian and other British colonial troops were engaged in combat in East Africa (Strachan 2004).

Although the Smuts–van Deventer strategy was intended to avoid decisive engagement and limit casualties, the results were very disappointing and in the end, resulted in high casualties anyway. Captain Richard Meinertzhagen, who was serving as a British intelligence officer from 1914 to 1916 was extremely critical of Smuts' and van Deveneter's strategy. He noted that their maneuver strategy backfired because of harsh environmental factors in the region. He noted in his diary that maneuver war, "*… may save casualties by bullet but involves hundreds on the sick-list in this pestilential climate and adds many millions on the war debt. It is the most extravagant form of warfare, and though successful in its initial stages against an inferior enemy, runs a good chance of utter failure in tropical Africa where climate and disease take such a heavy toll of men and animals, not to mention mechanical transport*" (Meinertzhagen 1960 p 195).

4.1 Terrain and Climate

Richard Dolbey served as a British medical officer in East Africa in 1916 and offered similar conclusions noting that, "*In this campaign the Hun has been the least of the malignant influences. More from fever and dysentery, from biting flies, from ticks and crawling beasts have we suffered than from the bullets of the enemy*" (Dolbey 1918 p 110). South African Deneys Reitz similarly wrote in his memoirs, "*The real enemy were the deadly climate, the wild regions, and the swamps and forests, and scrub*" (Reitz 1933 p 152). Arguably, East Africa proved to be the most physically challenging environment of the First World War. In his analysis of the military geography of the Western Front, Johnson (1921, 25) observed that, "… [he] *has no hesitation in saying that, of all the combatants, those who fought on the plain of Flanders endured the most terrible physical conditions*." In contrast, Dolbey (1918), who served in Flanders and East Africa, observed that, comparatively, the soldier in East Africa operated in a much more physically demanding environment, and subsequently suffered greater depredations resulting from the harsh operational environment.

Histories of the First World War convey a sense of the difficulties faced in East Africa and the significant influence of geography on its outcome. Nevertheless, published works highlight rather different scales of these affects. Primary accounts from diaries and memoirs articulate the human experience and emphasize the influence of geographic factors at the scale of the individual soldier. Memoirs of senior commanders and staff officers reflect these environmental challenges as well, but at much broader scales. Smuts' (1919, 442) reflections on the war are illustrative of the role of geography as he wrote, "*the difficulties and requirements of a large force, moving forward against an alert, ubiquitous foe, compel you to probe into everything: the nature of the country, with its mountains and rivers, forests and deserts, for scores of miles around; its animal and human diseases; its capacity for supplies and transport; its climate and soil and rainfall.*" As described above, von Lettow-Vorbeck's forces, commanded by seasoned officers, with several years' experience in the colonies provided German units with a distinct geographic intelligence advantage over their opponents.

Historical analyses of the campaign draw heavily from primary works and official histories to enable an understanding of the operating environment. The conflict in East Africa spanned a climatic gradient that ranged from coastal plains with a tropical climate (i.e., high temperatures, high humidity, and seasonally concentrated rainfall) to interior highlands (i.e., high temperatures and seasonal rainfall) (Sibley 1971; Miller 1974; Anderson 2004). Each region presented unique challenges to soldiers and different operational problems to leaders. Although publications impart a rather vivid imagery of the landscape and climate, most are rather broad, generalized descriptions with imprecise spatial references and thus provide little quantitative climatic data. Furthermore, most of the accompanying maps are simple planimetric diagrams that provide only a general concept of the terrain.

Fortunately, we can supplement our understanding of historical climate conditions in East Africa. Recent efforts to understand climate change has led scientists to collect and analyze historical climate data. Rubel and Kottek (2010) compiled and depicted climate zones for the last century using the Köppen-Geiger classification system, and data for the period between 1901 and 1925 provide the most precise climate approximations for this region. These data have been made available in shapefile format for use in ArcGIS (Fig. 2), and allow a much clearer understanding of the range of climates within which the conflict was fought.

The data given in Fig. 2 suggest that the majority of the conflict occurred in a tropical savanna (i.e., *Aw* or Equatorial Humid, dry winter) climate. In this type of climate region, the coolest month exceeds 18 °C (64.4 °F) and it has a dry winter (i.e., April to September). Specific temperature and precipitation data for particular places within this climate region during the period of the war are not available. However, proxy data have been compiled by Weatherbase (2011). Data for Africa have been drawn from averages during last 25 years and therefore do not provide an essentially accurate analysis of climatic conditions during the period of the war. However, despite this limitation, these data can serve as a useful approximation of historical conditions.

Weatherbase (2011) data were used to develop climographs for several locations in former German East Africa. Fig. 2 shows the climographs for Dar es Salaam, on the coast, as well as Tabora in the interior of the country. Both locations experience Tropical Savanna climates, with distinct dry periods (i.e., June through October). In Tabora, recorded annual temperatures average 23 °C (73.4 °F) with an annual mean high temperature of 29 °C (85 °F) and a low of 17 °C (62.6 °F). In coastal Dar es Salaam, annual temperatures average 26 °C (78.8 °F) with annual mean high temperatures of 29 °C (84 °F) and mean low of 22 °C (72 °F). The data also indicate that the amount of rainfall between November and April was such that it had a

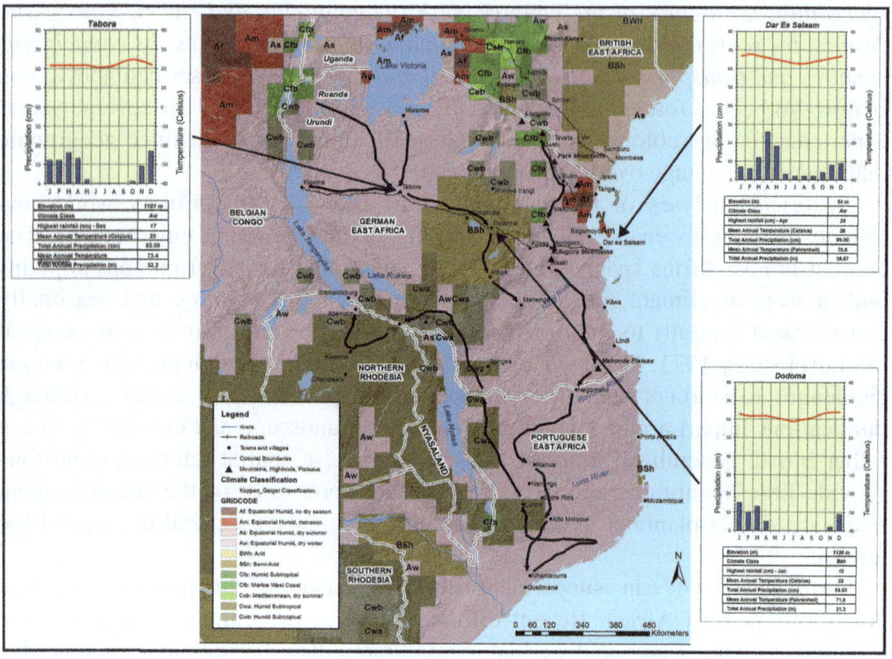

Fig. 2 Climate regimes and climographs for selected locations in East Africa. The climate zones map is drawn from Rubel and Kottek (2010) while climographs were constructed from historical data from Weatherbase (2011)

significant effect on soldiers and operational plans. Both regions averaged over 80 cm (30 in) of precipitation each year, although coastal areas experienced the highest levels of rainfall in April and May. The interior of the country received much greater rainfall, about 12 cm (5 in), in December through April.

Personal accounts describe in detail the effect these hot and humid conditions had on soldier morale and health. Christensen (2003 p 198) recounted Danish sailor Christen Noch's experience, "*and the rain came down as if the clouds shared our mood. There were single dry hours in the course of that first day, those days, those weeks, but rain was the normal thing. It soaked our clothes through and through, for we have not protection against it; it made dry fuel for fires at night almost impossible to find; it made small lakes in hollows of the forest and the plains; it filled dried-up water-course with brown liquid mud that swirled away down to the Rufiji, or towards the south.*"

This account reflects the effect that rain had on the individual, but also illustrates many other aspects of how the climate contributed to the overall difficulties in the campaign. While the lack of roads and limited rail lines restricted troop movements and resupply efforts, the rains exacerbated these difficulties by flooding rivers and low–lying areas, making roads impassable (Anderson 2004). The British depended more heavily on mechanized transport and subsequently many of their intended operations were affected by the onset of the rainy season. While several authors (i.e., Sibley 1971; Anderson 2004; Paice 2010) noted the effect that the rainy season had on operations, they do not discuss these patterns with sufficient detail to provide readers with a full understanding of when and where such conditions were prevalent. Although the data accessed from Weatherbase (2011) does not capture specific climatic conditions during the war, they do provide an approximate measure for understanding when and where these factors presented operational challenges.

As Fig. 2 also suggests, much of the conflict occurred in subtropical steppe or low-latitude hot desert climate regions and presented different operational challenges for combatants. The climograph for Dodoma (Fig. 2) reveals the annual temperature and precipitation levels in this region. Although Dodoma also experiences a rainy season (i.e., December through March) it received much less rainfall with relatively higher temperatures (annual mean 26 °C (78.8 °F)). In this regions, water supply was a much larger concern for both sides, and in many instances, water holes became key terrain (Meinertzhagen 1960). Particularly in the dry months, access to sufficient drinking water often became an overriding strategic imperative (Miller 1974).

4.2 Diseases

In addition to the challenges presented by the region's climate, the operational environment was also host to a variety of diseases, which ultimately caused large numbers of casualties. Anderson (2004) suggests that 20–25 percent of the soldiers were incapacitated by disease, and in some cases, far higher percentages of individual units were incapacitated by malaria, sleeping sickness, bilharzia, blackwater fever, dysentery. During the final stages of the conflict, the influenza pandemic reached the region (Strachan 2004; Sibley 1971). Nock noted, "…[malaria] *was so much more deadly than the English*" (Christensen 2003 p 187). Malaria was endemic throughout the region, but as Smuts (1919) wrote, "*The malaria mosquito is everywhere except on the higher plateaus.*" Other analyses indicate that it was much more prevalent in low–lying coastal areas and along the rivers and that units there suffered higher rates of debilitation (Sibley 1971; Miller 1974; Paice 2010). The higher elevations were considered healthier, and mountainous areas and plateaus therefore became key terrain (Strachan 2004). Note in Fig. 1 that the line–of–march used by von Lettow-Vorbeck's forces in the eastern region of the colony traversed areas of higher elevation in the north (i.e., Mount Kilimanjaro, Pare and Usambara Mountains) to the Nguru then Uluguru mountains before reaching the Makonde Plateau and then subsequently crossing the Rovuma River into Portuguese territory.

To safeguard their forces, quinine became a critical resource for both sides to protect against malaria. In this case British forces enjoyed the advantage of controlling ports, and were able to receive supplies of this and other needed medical supplies. For the Germans however, quinine supplies were quickly depleted. To limit the effects of malaria, and maintain forces in the field, they continued to acquire supplies through captured British stores, as well as making their own through boiling bark from cinchona trees (Miller 1974). Malaria was certainly a significant drain on both sides, but most analysts suggest that it had a much greater impact on Europeans (Dolbey 1918; Sibley 1971; Anderson 2004). Malaria was a particularly problematic factor for the British because, especially early in the war, their forces consisted of mostly British, Indian, and South African troops. Later however, they began to incorporate African recruits and the impacts of malaria declined somewhat. The much smaller number of German officers and NCOs in the *Schutztruppen* therefore required significantly smaller quantities of quinine to protect against the disease.

While malaria incapacitated and killed more soldiers than combat, the most prevalent disease affecting both sides was dysentery (Strachan 2004). The environmental conditions, constant movement, poor sanitation, and inadequate health care facilities contributed greatly to sickness rates. Limited medical supplies and inability to quickly resupply units further contributed to the prevalence of this condition as well as the evacuation of casualties to allow treatment and recovery. Poor roads and limited rail resources in East Africa forced both sides to rely heavily on animal transport. Yet here too, disease had a significant impact as Smuts (1919, 547) noted, "*everywhere the belts are infested with the deadly tsetse fly, which makes an end of all animal transport*." As the vector for trypanosomiasis (i.e., African Sleeping Sickness), the tsetse fly had a devastating effect on horses and mules, and during the course of the conflict, an estimated 150,000 animals were killed by this disease (Strachan 2004).

Although most publications describe the impacts of this disease, there are again few specific details provided within the historical literature on where these affected regions were located. Broad statements convey the sense that the entire region was plagued by the tsetse, yet cavalry and animal transport was used widely in some regions. The most often referenced sources for the distribution of tsetse–prone areas in Africa is Ford and Katondo (1977). Yet the Animal Health and Production Division of the Food and Agriculture Organization (FAO) of the United Nations has initiated a Programme Against African Trypanosomiasis (PAAT), building upon this work to not just depict regions of prevalence or absence of tsetse flies, but to identify and predict areas suitable for the infestation of tsetse. Analyzing vegetation cover, rainfall, temperature and elevation conditions through satellite imagery spatial analysis techniques, the Environmental Research Group Oxford (ERGO 2011) has provided predictability models for such areas for the three major species of tsetse in Africa: the fusca group (or forest tsetse), palpalis group (or riverine tsetse), and the morsitans group (or savanna tsetse). The results of these models are available in ArcGIS shapefile format. Although these models are based on more recent data, they provide a means by which to visualize areas where the different species of tsetse are (and were) likely to be found in Africa. Fig. 3 depicts the data compiled by the PAAT. As this figure illustrates, Riverine Tsetse offered little threat within the area of operations, but areas of much greater probability of tsetse infestation were

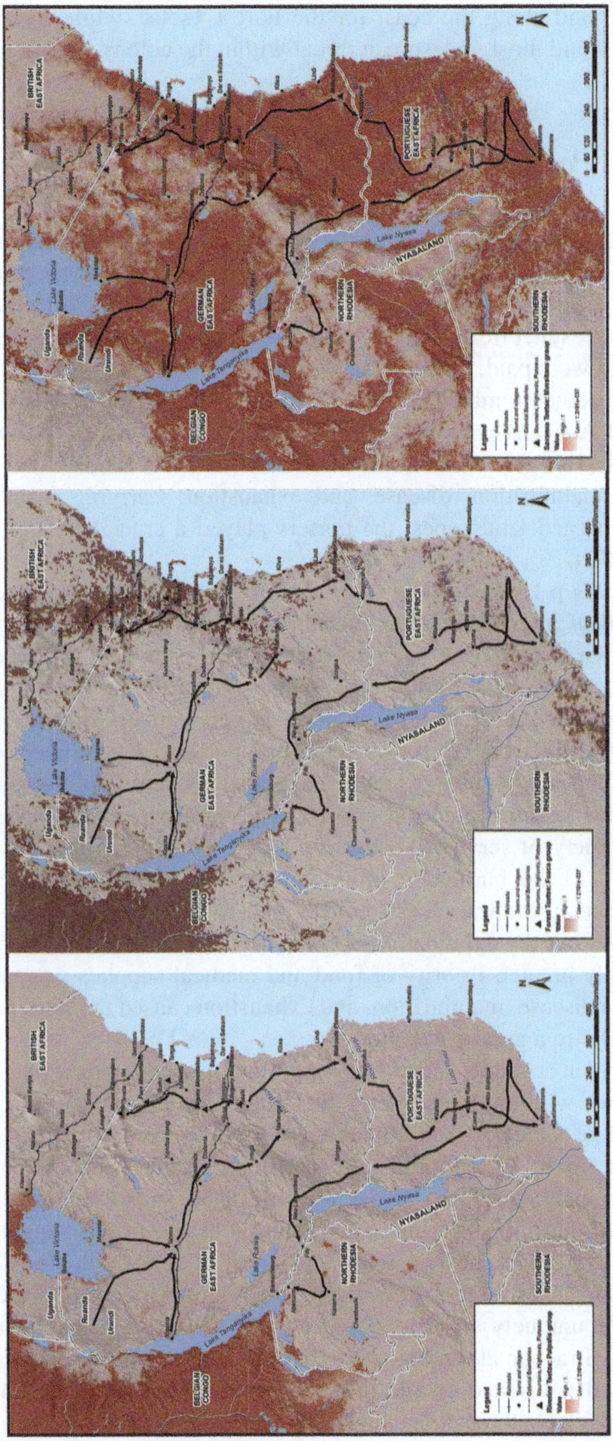

Fig. 3 Areas of high probability of Riverine, Forest, and Savanna Tsetse in East Africa. Data from Environmental Research Group Oxford (ERGO Ltd) (2011)

in the northeast and along the coast for the Forest Tsetse, while Savanna Tsetse posed the largest and most widespread threat within the colony.

4.3 African Porters, Carriers, and Civilians

As indicated in preceding sections, the natural and cultural landscape had a significant effect on the ability of the combatants to supply their forces. These factors presented extensive challenges, especially for the British, to feed and supply their forces. As a result, efforts to move equipment and supplies fell on the shoulders of Africans, who were paid, coerced, or conscripted to serve as porters and laborers to both armies. Actual numbers will likely never be known, but it is estimated that over one million Africans eventually served as porters for British and German forces (Strachan 2004; Hodges 1978). Of these, an estimated 100,000 or more died resulting from malnutrition, disease, and exhaustion. Carrying loads of 50–60 pounds across rugged landscapes, the porters played a crucial, though agonizing role in this conflict.

Anderson (2004 p 187) notes, "*a single Ford car of 135 kg cargo capacity could match the efforts of 300 carriers*," yet the nature of the terrain and the limitations of trucks in this environment necessitated large numbers of porters to support the war of maneuver. As the British continued to advance into German East Africa, supply lines became longer, which in turn required larger numbers of porters (Hodges 1978). In addition to those actively serving as porters, the number of Africans participating in the campaign was greatly expanded because families of the porters and *Askaris* often accompanied the units. Many contributed to the war effort by performing a variety of services such as cooking, gathering wood and water, but they also added to the burden of units already taxed by logistical constraints. Firsthand accounts of the war clearly convey the struggles of living and fighting in this environment, but the depredations suffered by the porters and camp followers were significantly greater. Priority of food and medical supplies was reserved for combatants; and disease, malnutrition, and exhaustion caused the porters and camp followers to suffer to a greater extent.

Although written accounts address the important role that porters played, there is a noticeable shift within contemporary literature. Earlier works (i.e., pre-1980s) address the size of the effort and impact the porters had on supporting combatant units. However, more recent publications offer expanded discussions on how indigenous Africans contributed to the war effort, but also how their lives were severely affected and the sufferings they endured as they were caught up in the European drive for colonial expansion. Page's (1987) edited volume in particular, addresses the impact of the Great War on the lives of Africans, by attempting to provide a more uniquely African perspective. Other recent publications (i.e., Anderson 2004; Strachan 2004; Paice 2010) focus significantly more discussion on the impact that the war had on the lives of African soldiers. Additionally, these publications underscore the fact that the war was not merely fought by Europeans in the dense ‘bush’ of Africa, but that the conflict raged across a vast landscape, directly or indirectly involving or affecting large numbers of Africans who had little or no stake in its outcome.

Like much of the literature in which the trenches of the Western Front figure prominently in images and reflections of the war, the campaign in East Africa is likewise constructed largely around combat fought in the bush. While most of the actual fighting occurred in such settings, towns and villages throughout the region were prominent objectives and key terrain. Ports were critical locations, but interior settlements were also similarly important for a number of reasons. As critical trail, road, and rail junctions, towns and villages served as important nodes during the war of maneuver. Just as important, villages and towns with buildings, houses, and standing structures provided a measure of protection from the elements, particularly for use as makeshift hospitals and causality collection points (Dolbey 1918). Despite the insidious unsanitary conditions found in many of these structures, they at least offered protection from the elements.

As the campaign dragged on, casualties became an increasing burden on the logistical system. Rather than leave wounded and sick behind in the bush, settlements became important sites in which those unable to travel further were left behind, either to await evacuation, or to be ‘surrendered’ to the opposing side in the hope they would receive sufficient care. Ludwig Deppe, a doctor serving in Lettow’s command, wrote in his memoirs, “*Behind us we leave destroyed fields, ransacked magazines, and for the immediate future, starvation. We are no longer the agents of culture; our track is marked by death, plundering and evacuated villages, just like the progress of our own and enemy armies in the Thirty Years’ War*” (Strachan 2004 p 95).

5 Conclusions

In his analysis of the American Revolution Palmer (1975 p 25) notes that, “*it becomes exceedingly difficult to fathom more than the sketchiest details of what happened, virtually impossible to know why, without a clear glimpse of the theater environment.*” The large body of literature on the East African campaign details the history of the war on this forgotten front, and provides a general understanding of its physical environment. This chapter examined the East east African campaign through the lens of the military geographer to provide a more thorough analysis and understanding of the geographic factors that were strategically decisive.

Although space limitations do not permit a comprehensive analysis using the complete environmental matrix, the focus on the most salient aspects of the physical and cultural landscape has hopefully shed light on two important aspects of military geography. First is the continued significance this perspective provides towards a greater understanding of how geography shapes the course of military operations. Clearly, as most accounts describe, the environment played a crucial role in the course of the East African campaign. However, more importantly, this operational environment also drove the need for colonial powers to rely on, and subsequently involve, hundreds of thousands of Africans. This second factor illuminates how civilians were drawn into combat, which can be explained by the region's geography. Increasingly, the impact on and roles of civilians on the battlefield are entering into tactical and strategic considerations and the topic of civil–military operations have assumed much greater prominence with current conflicts in Iraq and Afghanistan. Analyzing how and why civilians have been caught up in combat operations can hopefully lead to better strategies to avoid such occurrences in the future.

The war in East Africa was driven by an imperialist vision and the desire of European powers to control access to vital natural resources in Africa. This chapter has also revealed the tragedy of this conflict. Although considered by some as one of the most successful examples of guerilla warfare and a brilliant tactical campaign on the part of the German commander, more recent analyses have questioned these conclusions and argue that the campaign did nothing to alter the course of the war. The war was decided on the Western Front, and Germany was forced to capitulate and accept the demands of the Allies, thereby losing all of her colonial possessions. This likely would have been the result regardless if von Lettow–Vorbeck had been able to defeat British forces and retained control over German East Africa at the end of the war. The end result was a meaningless campaign that ultimately changed nothing, except to disrupt and devastate the lives of so many Africans who had little care for war between foreigners so far from their homes.

Literature Cited

Abbott P (2002) Armies in East Africa. Osprey Publishing, Oxford, pp 1914–1918

Anderson R (2004) The Forgotten Front: The East African Campaign 1914–1918. Tempus Publishing Limited, Wiltshire

Cann JP (2001) Mozambique, German East Africa, and the great war. Small Wars Insurg 12(1):114–143

Chasseaud P (2002) British, French, German mapping and survey on the Western front in the first world war. In: Doyle P, Benn MP (eds) Fields of battle: terrain in military history. Kluwer Academic Publishers, Boston

Christensen CP (2003) Blockade and jungle. Battery Press, Nashville

Dolbey RV (1918) Sketches of the East African campaign. J Murray, London. http://www.gutenberg.org/ebooks/10362

Doyle P, Bennett M (eds) (2002) Fields of battle: terrain in military history. Kluwer Academic Publishers, Boston

Ellis J (1976) Eye–deep in hell: trench warfare in world war I. Johns Hopkins University Press, Baltimore

ERGO (Environmental Research Group Oxford) (2011) Programme Against Trypanosomosis. Food and Agriculture Organization of the United Nations. http://www.fao.org/ag/againfo/programmes/en/paat/maps.html. Accessed 12 Feb 2012

Farwell B (1986) The great war in Africa. W.W. Norton and Company, New York, pp 1914–1918

Flint C (2006) Introduction to geopolitics. Routledge, New York

Ford J, Katondo KM (1977) The distribution of tsetse flies in Africa. Cook, Hammond & Kell, Nairobi

Galgano FA (1998) The environmental matrix. In: Galgano FA (ed) Readings in military geography. Department of Geography and Environmental Engineering, United States Military Academy, West Point

Garver J (1975) The environmental matrix. In: Garver J (ed) Readings in military geography. Department of Geography and Environmental Engineering, United States Military Academy, West Point

Hodges GWT (1978) African manpower statistics for the British forces in East Africa, 1914–1918. J Afr Hist 19(1):101–116

Homer–Dixon TF (1991) On the threshold: environmental changes as causes of acute conflict. Int Secur (2):76–116

Hoyt EP (1981) Guerilla: Colonel von Lettow-Vorbeck and Germany's East African empire. Schribner, New York

Johnson DW (1921) Battlefields of the world war, Western and Southern fronts: a study in military geography. Oxford University Press, London

Keegan J (1998) The first world war. Knopf, New York

Keithly D (2001) Khaki foxes: the East Africa Korps. Small Wars Insurg 12(1):166–185

Meinertzhagen R (1960) Army diary. Oliver and Boyd, London, pp 1899–1926

Meyer GJ (2006) A world undone. Bantam Dell, New York

Miller C (1974) Battle for the Bundu. McMillan, New York

Ofchansky T (1990) Introduction. In: von Lettow–Vorbeck P (ed) My reminiscences of East Africa. Battery Press, Nashville

Page M (ed) (1987) Africa and the first world war. St Martin's Press, New York

Paice E (2010) World war I: the African front. An imperial war on the African continent. Pegasus Books, New York

Palka EJ (2004) Military geography: its revival and prospectus. In: Gaile GL, Willmott CJ (eds) Geography in America at the dawn of the 21st century. Oxford University Press, London

Palmer DR (1975) The way of the fox: American strategy in the war for America. Greenwood Press, Westport, pp 1775–1783

Reitz D (1933) Trekking on. Faber and Faber, London

Rubel F, Kottek M (2010) Observed and projected climate shifts 1901–2100 depicted by world maps of the köppen-geiger climate classification. Meteorol Z 19:135–141

Sibley JR (1971) Tanganyikan guerrilla: East African campaign 1914–18. Ballantine Books, Inc., New York

Smuts JC (1919) East Africa. In Reynolds FJ, Churchill AL (eds) World's war events, volume III, eds., Project Gutenberg EBook. http://www.gutenberg.org/files/16513/16513-h/16513-h.htm. Accessed 1 Feb 2012

Strachan H (2004) The first world war in Africa. Oxford University Press, New York

von Lettow-Vorbeck P (1990) My reminiscences of East Africa. Battery Press, Nashville

Weatherbase (2011) http://www.weatherbase.com. Accessed 2 Oct 2011

Winters HA (1998) Battling the elements: weather and terrain in the conduct of war. The Johns Hopkins University Press, Baltimore

Conflict in the Horn of Africa: The Ogaden War of 1977

Francis A. Galgano

Abstract The nexus of environmental insecurity, disasters, and conflict have become an essential paradigm in security planning, policy, and analyses. The U.S. National Intelligence Council warns that the likelihood of environmentally triggered conflict will increase in the coming decades. Nonetheless, many scholars dismiss this outlook. History appears to support their position because these problems have typically been resolved using peaceful, diplomatic or economic means. Furthermore, it is difficult to establish clear cause–and–effect links between disasters, environmental stress, and armed conflict. However, the security landscape has changed decisively. This chapter suggests that continued peaceful resolution of potential conflicts with an environmental component is incongruous with the realities of the emerging national security landscape. First, climate change and demographic factors are degrading environments and magnifying the effects of environmental degradation and resource shortages beyond the management capacity of many states. Second, the proliferation of failing states has singularly reduced the potential for diplomatic resolution in many regions. Finally, competition for essential resources has been intensified by population growth. Thus, I argue that environmental factors will likely provide a tipping point for regions that already manifest severe environmental degradation and civil unrest; and these insidious problems can be exacerbated by disasters or other short–term climate shocks. The 1977 Ogaden War is one such example and is used as a case study to illustrate these dynamics. An analytical framework is used to illustrate the factors of environmental change, non–sustainable practices, human activity, and governance in Ethiopia during the 1970s to demonstrate their role in triggering the Ogaden War.

Keywords Adaptation · Climate change · Climate change · Cold war · Conflict · Desertification · Drought · Environment–conflict Nexus · Ethiopia · Ethnic warfare · Exposure · Governance · Migration · Ogaden · Ogaden war · Red terror · Somalia · Vulnerability

---

F. A. Galgano (✉)
Department of Geography and the Environment, Villanova University, Villanova, PA, USA
e-mail: francis.galgano@villanova.edu

F. Galgano (ed.), *The Environment-Conflict Nexus*, Advances in Military Geosciences, https://doi.org/10.1007/978-3-319-90975-2_9

1 Introduction

The 1977 Ogaden War is a particularly instructive environmental security case study because it is a very clear example of the environment–conflict nexus. All wars are insidious and destructive, but in this case, the conflict was coincident with a devastating drought and a series of anthropogenically–induced environmental disasters, which intensified extant political and societal instability in Ethiopia (Westad 2007). This is notable because contemporary research suggests that the frequency of climate–related disasters is increasing (Guha–Sapir et al. 2004; Smith and Vivekananda 2009; Trondalen 2009; IPCC 2012) and that the prospect of violent conflict, triggered by environmental instability, is certainly plausible (Kaplan 2000; Diehl and Gleditsch 2001; Smith and Vivekananda 2007; Reuveny 2007; Burke et al. 2009; Hsiang et al. 2011; Solow 2011; Hendrix and Salehyan 2012). Hence, the Ogaden War is an excellent example of the environmental security paradigm, which has emerged as one basis for understanding violent conflict because it demonstrates linkages between instability induced by environmental stress (e.g., disasters, climate change, and anthropogenic factors), governance, civil unrest, demographic factors, and warfare. During the 1970s, the Ethiopian highlands experienced persistent drought and widespread famine, which were exacerbated by non–sustainable practices, population growth, and political instability. The drought and famine elicited a mass migration into the Ogaden region (Fig. 1), against which Somalia maintained irredentist claims.

Somalia invaded the Ogaden region of Ethiopia on 23 July 1977, thus initiating a short, virtually unknown war that led to pervasive instability on the Horn of Africa that persists to this day (Westad 2007). The Ogaden War lasted nine months, and although casualty figures are variable, conservative estimates suggest that it resulted in about 37,000 military dead and perhaps 65,000 civilian dead from direct military action (Tareke 2000). The number of people injured and displaced by the war is thought to be in the tens of thousands. The war was clearly a humanitarian disaster, however, its scope was magnified because at the same time Ethiopia was experiencing a decade–long drought and famine, which killed an estimated two million people and caused the mass migration of another three million (Westing 1991; Nkaisserry 1997; Guha–Sapir et al. 2004).

This chapter examines the human and environmental conditions in Ethiopia that contributed to the Ogaden War. This case study does not suggest that environmental stress is a sole cause of warfare. To be more precise, it can potentially trigger violent conflict in unique situations of extreme civil instability and environmental stress within failing states as was the case in the Horn of Africa in 1977 (Westing 1991). The problem that we face today is that the number of failing states is growing (Galgano 2007), and they are more vulnerable to instability caused by environmental stress because they suffer from four related effects: (1) diminished agricultural production and food insecurity; (2) economic decline; (3) population displacement; and (4) civil disruption (Homer–Dixon 1991; Porter 1995; Lee 1997; Reuveny 2007). Furthermore, a growing body of evidence suggests that the incidence of

Fig. 1 Map of Ethiopia depicting the main Somali attack in July–December 1977. Cartography by the Author

environmental and climate–related disasters is on the rise, and the number of people affected is increasing (IPCC 2012, 2014). This study uses an environmental security framework proposed by Krakowka et al. (2012) to analyze the complex and interrelated human and environmental factors that triggered the Ogaden War.

2 Disasters, Environmental Instability, and the Security Landscape

The contemporary world is facing environmental disasters on an unprecedented scale: one in twenty-five people worldwide have been affected by disasters since 1994 and they have claimed more than 58,000 lives annually. Many researchers attribute this spike to rapidly growing populations and the pervasive effects of global climate change, which may be intensifying natural disasters such as drought, storms, and floods (Guha–Sapir et al. 2004; IPCC 2012). Geography, as an integrating spatial discipline, provides a particularly effective vantage point from which to examine climate–induced disasters, and geographers are widely recognized as founders of the discipline of disaster and hazard analysis (Cutter 1993). Geographers, too, have had a profound influence on our general understanding of the full scope of crises generated by complex interactions of natural and human processes (Cutter et al. 2002). Understanding and explaining these interactions are essential because we tend to classify disasters into discrete categories such as natural and anthropogenic; yet in reality these divisions are not clear–cut. In fact, most disaster researchers argue that it is not possible to make well–defined distinctions between these classifications because many disasters—floods and famine by way of example—are often triggered by the interaction of a number of natural and anthropogenic factors (Alexander 1993).

The situation in Ethiopia during the early 1970s highlights the important and complex interactions of natural and anthropogenically–generated disasters, human processes, and the security landscape. Persistent drought in Ethiopia begat famine, however, the famine was intensified by human–induced desertification, non–sustainable agricultural practices, a lack of infrastructure, and ineffective governance (Myers 2004). The insidious and reinforcing effects of these natural and anthropogenic factors created a highly stressed environment, which clearly contributed to the destabilization of Ethiopia and set the stage for the Somali invasion of the Ogaden (Westad 2007). Events such as the Ogaden War imply that a nexus exists between disasters, environmental stress, and conflict because these stresses intensify latent ethnic divisions, poor governance, and social stratification, thus further undermining stability beyond the adaptation ability of the state (Homer–Dixon 1999). Evidence suggests that this trend will persist because climate change and population growth will continue to stress marginal environments in places with inherently weak governance (Smith and Vivekananda 2007; IPCC 2012).

3 The Environment–Conflict Nexus

Environmental security refers to a broad range of security issues intensified by environmental stress, resource shortages, and demographic factors (Galgano and Krakowka 2010). Environmental security is perhaps the most transnational of transnational issues because it does not respect sovereign boundaries and it tests our understanding of national security processes. The environmental security model is dependent on four critical variables: environmental stress, resource scarcity, governance, and political/societal response (Galgano and Krakowka 2010). Environmental stress is caused by climate change, disasters, and non–sustainable practices. Scarcity is determined by consumption, population change, and the unequal distribution of resources. Governance essentially determines the ability of civil authorities to adapt to a stressed environment. This is important because by the end of this century, about one–third of the Earth's surface will be significantly altered and its population will approach nine–billion (Floyd 2014). Consequently, understanding the implications of the environmental security concept are important because conflicts with an environmental component coupled with divisive ethnic/political dimensions, such as those observed in the Ogaden case study, have increased pressure on the West and U.N. to commit resources to stability and disaster relief efforts (Klare 2002; Guha–Sapir et al. 2004).

Not all environmental problems lead to violent conflict and not all conflicts emanate from environmental stress. In fact, it is rare for linkages between environmental stress and conflict to be directly and exclusively causative because there are many other variables mixed in, such as ineffective economies, repressive governments, ethnic conflicts, and social stratification. However, environmental security is not simply a neo-Malthusian paradigm (Myers 2004). The environment can play a decisive role in triggering conflict, and environmental security doctrine is only one plausible explanation, it is not a deterministic concept (Porter 1995). Scarcity and stress contribute to conflict only under particular conditions, but there is no deterministic link (Homer–Dixon 1999). Dynamics between environmental stress and conflict are complex and the outcome of a potential environmental security scenario is influenced strongly by government policy, social structure, governance, technology, and infrastructure (Mathews 1989; Solow 2011).

Contemporary research has, however, defined clear links between environmental stress and conflict. Hsiang et al. (2011) developed a quantitative model using ENSO data from 1950–2004 and demonstrated that the probability of conflicts doubles in the tropics during El Niño years. Burke et al. (2009) conducted a comprehensive examination of global climate change and its potential linkages to armed conflict in sub-Saharan Africa, and suggest that there will be a 54% increase in the incidence of armed conflict by 2030. Hendrix and Salehyan (2012) examined deviations from normal rainfall patterns in Africa and their results indicate that extreme variations are associated positively with political and civil conflict. Thus, it appears that environmental change and scarcity is already contributing to instability and violence in the developing world (Solow 2011). The critical problem is defining the tipping point between a highly degraded and stressed environment, and those societies and governments that can't adapt (Homer–Dixon 1999). The real problem facing the West is that, in the developing world, the capacity to adapt is declining as governments continue to fail (Galgano 2007).

4 The 1977 Ogaden War

The Ogaden War began in July 1977 when Somali forces crossed into Ethiopia to bring to fruition its expansionist dream of a *Greater Somalia*. This confrontation, which became one of the classic proxy conflicts of the Cold War era, lasted for 9 months, and is one of the two largest wars between African states in contemporary times (Nkaisserry 1997). The Somalis were eventually defeated and the war cost each side approximately 35,000 dead along with an estimated 2000 Cuban mercenaries (Tareke 2000). The total human toll in civilian dead, displaced people and disrupted lives possibly numbers in the millions, but likely will never be known (Westing 1991). Nonetheless, the Ogaden War clearly illustrates that armed conflict is one of the greatest disasters that can befall a population, and that its effects linger for many decades afterward. The Ogaden War too, illustrates the fact that in modern warfare, civilians now bear the majority of injuries and deaths (Christenson 2008). The war effectively ended Somalia's ability to function as a state and the entire region was plunged into three decades of instability and conflict, which persists in the Horn of Africa to this day.

4.1 A Military Geography of the War

The Ogaden War began on 23 July 1977 and was fought in two distinct phases. The first, the Somali offensive, lasted until about December when their attack was halted on the outskirts of Dire Dawa (Fig. 1). The second phase began in early 1978 with an Ethiopian counter–offensive, which ultimately drove the Somalis from Ethiopian territory. The conduct of the war was tempered by the timeless geographic factors of time and space, terrain, key transportation nodes, and climate and weather (Tareke 2000). The frontier between the two states is about 900 miles long, but the principal military effort was concentrated along a very narrow corridor extending from the main Somali base at Hargeysa to Dire Dawa (Fig. 1) (Harkavy and Neuman 2001). The Somali strategy had two mutually supporting objectives. First, they planned to use regular forces combined with irregular units, composed of ethnic Somalis (i.e., the Western Ogaden Liberation Front), to overrun the large, sparsely populated semiarid upland plateau of the Ogaden Desert in the general direction of Werder (Fig. 1). The second, and decisive element of their strategy, was their main attack toward the heartland of Ethiopia along the axis Hargeysa–Jijiga–Harar–Dire Dawa (Fig. 1). This drive was intended to sever Ethiopia's main rail link to the Red Sea, which could have forced the total defeat of Ethiopia, or at least forced it to cede the Ogaden region (Harkavy and Neuman 2001).

The war began well for the Somalis, and by the end of July, they had captured about 60% of the Ogaden largely using irregular forces, which seized small, widely spaced Ethiopian garrisons on the Ogaden plateau. Along the decisive axis of the war, the Somalis captured Jijiga and seized the key upland passes as they approached Harar and Dire Dawa. However, the offensive stalled in front of Harar as the Somalis became overextended and lost their ability to maneuver in the rugged mountains (Tareke 2000; Schwab 1978). Here, Ethiopian forces, reinforced with Cuban mercenaries and Soviet weapons, eventually stopped the Somalis in the highlands along the southern rim of the Great Rift Valley. In February 1978, the Ethiopians began their counteroffensive along two principal axes: the main axis east from Harar toward Jijiga, and a secondary attack north from Dire Dawa toward Djibouti (Westad 2007). The main thrust toward Jijiga finally succeeded on 5 March, and Ethiopian forces reached the border with Somalia on 10 March 1978. On 15 March, both sides agreed to a truce, brokered by the U.S. and U.S.S.R., which essentially restored the *status quo* (Tareke 2000).

The Ogaden conflict demonstrated the futility of a preemptive war initiated by a smaller, weaker state against a larger, more powerful opponent. The Somalis viewed

Ethiopia as greatly weakened and they opted to seize a temporary advantage and attacked into the Ogaden (Tareke 2000). However, geography clearly worked against them. The Ethiopians adroitly used their great strategic depth by not engaging in a decisive battle with the Somalis, and by the winter, the factors of terrain and distance stopped the invaders in front of Harar as much as any military effort by Ethiopian forces (Harkavy and Neuman 2001). During their initial advance on Jijiga, the relatively open terrain enabled the Somalis to fight a war of maneuver, but that advantage was lost once they entered the constricted, mountainous terrain in front of Harar and Dire Dawa (Schwab 1978).

The Ogaden War has its roots in three important events. First, Ethiopia's monarch, Haile Sellassie, was deposed by a military *coup–de–etat* in 1974. The Mengistu government, which followed, proved to be ineffective, and the country devolved into a near civil war. Thus, the deteriorating political and economic situation in Ethiopia convinced the Somalis that the time was right to invade the Ogaden. Somalia's irredentist claims were the second driving factor behind the war. Following their independence in 1960, the Somalis pursued a nationalist agenda that included long–held irredentist claims on the Ogaden as well as territory in Kenya and Eretria (Ofcansky and Berry 1991). Thus, there were important politico–military factors that drove Ethiopia and Somalia into a military confrontation. However, in this case there is a third—environmental—dimension that must be taken into account. The crushing drought, which affected the region beginning in 1969, caused widespread famine and exacerbated pressures created by civil instability within Ethiopia to the extent that it caused the forced migration of nearly three million people into the Ogaden (Nkaisserry 1997).

4.2 An Environmental Security Analysis

The Ogaden War illustrates many of the elements of the environment–conflict nexus: i.e., a failing state with ethnic conflict and social stratification, rapid population growth and mass migration, environmental degradation, and resource competition. In the environmental security model, these factors combine to exceed the adaptation ability of the government, which may eventually lead to violent conflict. The Ogaden War represents a historically destructive convergence of natural and anthropogenic disasters that overwhelmed the entire region—but Ethiopia in particular—and serves as a useful example of the environmental security paradigm. The region experienced a drought that lasted for more than a decade. In Ethiopia, the drought combined with non–sustainable land use and agricultural practices to initiate a massive famine that was exacerbated by the devolution of the state, which essentially meant that survival for most Ethiopians was a matter of their own personal initiative (Khalif 2000). Guha–Sapir et al. (2004) estimate that between 1974 and 1977 the total number of Ethiopians killed or otherwise adversely affected by the drought and famine was about 1.9 million people. They further estimate that in the decade that followed the war, another 22 million Ethiopians

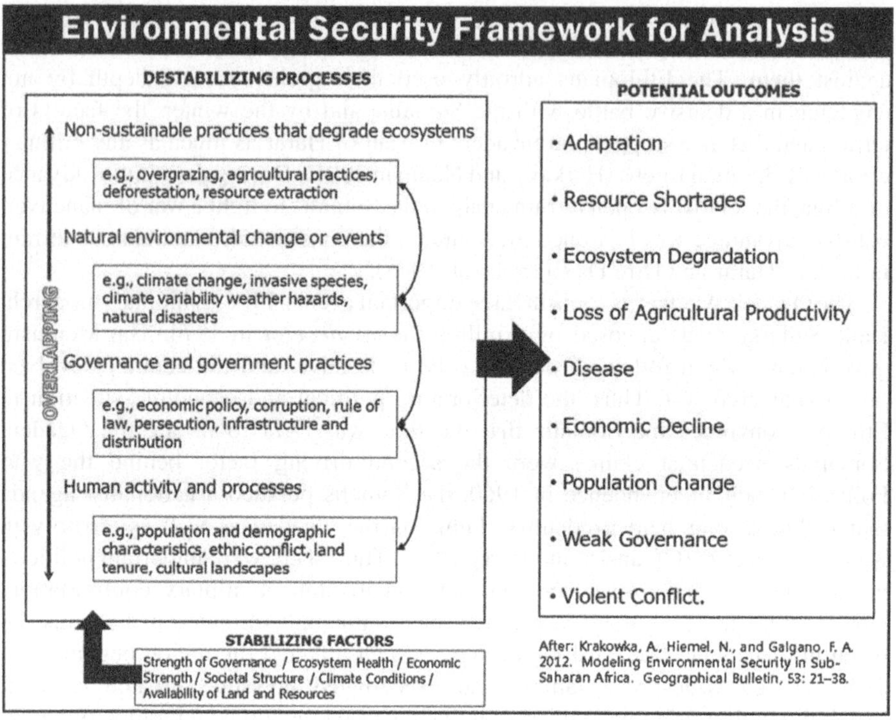

Fig. 2 Framework for environmental security analysis (Krakowka et al. 2012)

were killed or adversely affected (Guha–Sapir et al. 2004). These numbers have to be considered with the effect of the drought and famine on the entire region. During the period 1974–1988, it is estimated that some 460,000 people were killed in Djibouti and another 1.1 million in Somalia (Guha–Sapir et al. 2004). Major climate disasters typically trigger or encourage mass migrations and this was certainly the case in Ethiopia (Christensen 2008). As the drought and famine deepened during the 1970s, population growth, poverty, and political instability compounded these problems, and initiated a mass migration of some three million people into the Ogaden Region, which Somalia viewed as an encroachment on its historical, sovereign territory (Khalif 2000; Tareke 2000).

Although the Ogaden case study offers a compelling example of environmental security, it also underscores the principal weakness of the model, which is a lack of predictive capacity. That is, we have no overarching sense of which environmental scenarios will lead to conflict in the future (Myers 2004). Therefore, an analytical framework is helpful to understanding the complex, and interrelated components of such a conflict. Krakowka et al. (2012) developed a framework (Fig. 2) to examine the nexus between environmental stress, instability, and conflict. Their investigation of environmental security in Sub–Saharan Africa demonstrated a need for careful analysis using such a framework to examine the complex relationships between degraded environments, natural environmental change, political stability, and human activity. The framework is not a checklist because they tend to disconnect intrinsically linked factors into artificially discrete variables. The elements of their framework are not intended to be independent or sequential; in fact, they take place concurrently. Furthermore, they are synergistic and amplify their mutual influence on the ecosystem and population. For that reason, the framework given in Fig. 2 does not imply a linear or sequential relationship that is characteristic of other environmental security models in the literature. This analytical framework (i.e., Fig. 2) is used to assess the relevant components of this case study.

4.2.1 Non–sustainable Practices

In 1977, Ethiopia was a rural society and the security and productivity of its people was entirely tied to the land as subsistence farmers and nomadic herders. They routinely suffered from the depredations attendant to periodic crop failures and loss of livestock caused by periods of drought or excessive rain and floods. As the population grew, farms encroached on forests and marginal lands (e.g., slopes, etc.), causing deforestation, accelerated soil erosion, and a significant lowering of the water table (Ofcansky and Berry 1991). Rangelands used by pastoral nomads became increasingly restricted and this led to a rapid denudation of the grasslands, which accelerated desertification in the highlands (Kalif 2000). These non–sustainable practices slowly diminished food supplies, and thus Ethiopia was already approaching the brink of a severe food crisis when the onset of an unprecedented drought in the late 1960s initiated a famine that persisted well into the 1980s. These conditions contributed to relentless civil and political instability, which ultimately led to the fall of Haile Sellassie and then the collapse of the Mengistu regime as both governments lost their legitimacy to rule (Westing 1991).

4.2.2 Natural Environmental Change

Clearly, the single most damaging environmental variable was the drought, which expanded eastward through the Horn of Africa, and by the middle of the 1970s, starvation threatened the lives of millions of Ethiopians (Westing 1991). On average, the Ethiopian Highlands receive approximately 1229 mm (~48 in.) of rain annually (NCDC 2013). The region has a grassland climate within which precipitation and evapotranspiration are essentially balanced. However, this is a fragile ecosystem that cannot withstand long departures from this equilibrium. At the onset of the drought (Fig. 3), average annual precipitation declined to approximately 970 mm (~38 in.); and by the middle of the decade, the region was only receiving about 600 mm (~23 in.) per year (Conway et al. 2004). Thus, by 1974, famine brought about by the drought had claimed the lives of about approximately 300,000 people (Nkaisserry 1997). After assuming power in 1974, the Mengistu regime embarked on a program to improve the condition of the peasants, but the famine persisted

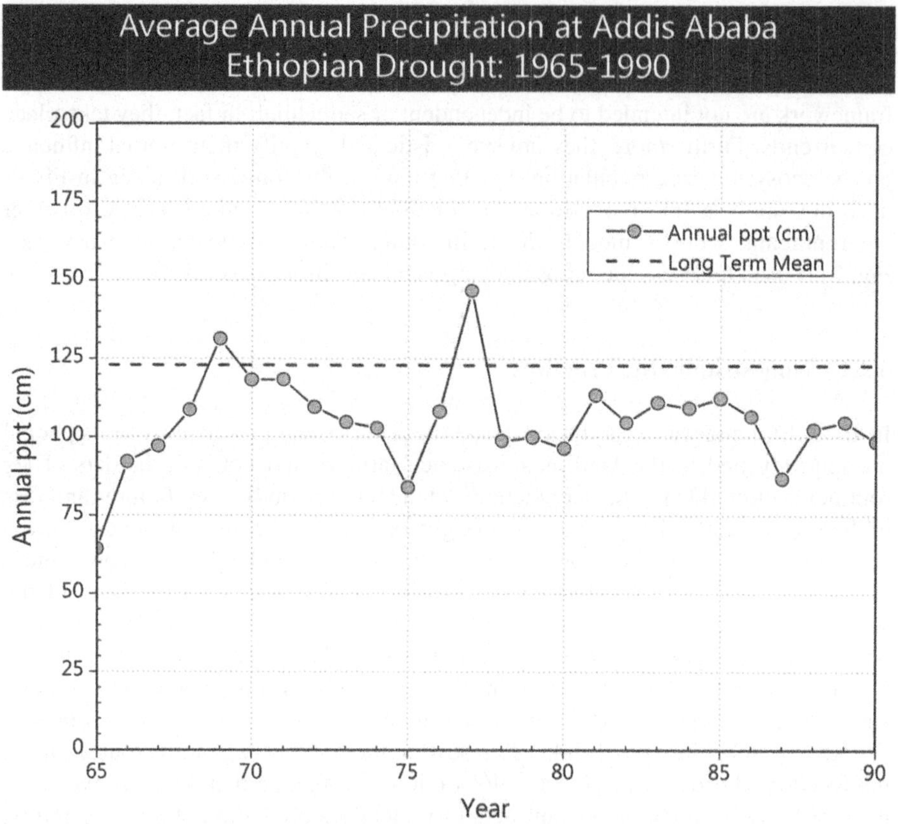

Fig. 3 Rainfall data for Addis Ababa, 1965–1990 (NCDC 2013)

despite these efforts; and by 1977, drought and famine forced more than three-million people to migrate into the Ogaden to seek relief (Westing 1991).

4.2.3 Governance and Government Practices

Following the Second World War, Ethiopia labored under the rule of a rigid, centralized monarchy. Haile Sellassie led a poorly developed country and modernization was accepted only so long as it enhanced the power and prestige of the monarchy (Nkaisserry 1997). The imperial elite that dominated the country actively blocked the adoption of Western technology as well as agricultural and industrial methods, and thus the country was composed almost entirely of subsistence farmers and nomadic pastoralists. By the time of the *coup* in 1974, there were only about 50,000 skilled workers in all of Ethiopia. Consequently, the country was technically unsuited to adapt to the crushing effects of the drought and famine (Westad 2007). Thus, Ethiopia was one of the poorest countries in the world with a

1975 GNP of $90 *per capita*; and, by every standard, it was a failed state and was rated among the twenty–five most impoverished in the world (Ofcansky and Berry 1991). Although Selassie was seen as the father of African independence, the Ethiopian people began to regret the lack of opportunity and extreme backwardness of rural areas. Corruption and general governmental inefficiency further eroded Selassie's ability to govern effectively. As the drought and famine began to affect a larger segment of their society, Ethiopians viewed the lack of a coordinated government response as evidence that the government had lost its legitimacy to rule (Westad 2007).

Yet, opposition to the regime by average citizens was not particularly strong. In the end, it was activism by junior officers in the army that eventually provided the impetus for change, and a military–led *coup* overthrew Selassie in 1974. The new military–led government, the *Derg,* took power, but it had very few followers and offered only a vague program for improvement (Westad 2007). This led to a period of shifting leadership and further chaos until the ascension to power of Major Mengistu Haile Mariam. Nevertheless, his government proved to be equally ineffective, and rather than propel Ethiopia on a path of modernization and recovery, the country devolved into economic, civil, and political chaos. For example, his attempt to promote land reform was widely accepted by the majority of Ethiopians, but breaking up local patterns of power, social customs, and land tenure practices proved to be very difficult and these efforts largely failed (Westad 2007).

The new regime also faced a difficult war in Eretria and growing insurgent movements in Tigray Provence and in the Ogaden, which only intensified the effects of the environmental disaster that the average citizen faced (Nkaisserry 1997). By 1977, Ethiopia was in a state of continuous internal conflict. In response, the Mengistu regime instituted the so–called "Red Terror" in 1977 during which government security forces killed thousands of students, engineers, teachers, and scientists (Tareke 2000). Thus, Ethiopia entered the spiral of environmental stress and violence that characterizes the environmental security paradigm. Developing states, such as Ethiopia, are more susceptible to this spiral of violence because they are, characteristically, more dependent on the environment for their economic productivity, and they do not have the institutional and economic resiliency to overcome exposure to adverse environmental effects (Galgano and Krakowka 2010).

Thus, the stage for the war was set. During the 1970s, Somalia was promoting a nationalist and expansionist agenda and also maintained strident irredentist claims against the Ogaden region within Ethiopia with its large ethnic Somali population (Tareke 2000). The Somali regime viewed the extreme instability within Ethiopia as a sign of weakness, but more importantly, the environmentally induced mass migration of millions of ethnic Ethiopians into the Ogaden may have triggered the Somalis to attack and attempt secure the region from this apparent encroachment (Nkaisserry 1997).

4.2.4 Human Activity and Processes

Population and demographics are fundamental variables in the environmental security calculus. Ethiopia's 1950 population was approximately 18.4 million people, and by 1975, it nearly doubled to about 32.9 million. This growth strained marginal environmental conditions and ineffective governmental institutions (Ofcansky and Berry 1991). Ethiopia was also confronted with significant demographic problems and social stratification, and the devolution of the state was propelled by ethnic tensions as well. Ethiopia's economic divisions coincided with its ethnic and social strata. There are three major ethnic groups of Ethiopians, the Amharas, Tigres, and Gallas (Ofcansky and Berry 1991). These ethnic Ethiopians dominated the state's political, civil, and economic institutions and the other ethnic groups that formed the lower strata of Ethiopian society, the ethnic Somalis and Eritreans in the northern and eastern parts of the country were alienated from Ethiopian society (Westad 2007). These groups were effectively excluded from participation in any of the institutions of the Ethiopian government and were economically marginalized, which fostered their anti–government violence (Tareke 2000). Consequently, they lived in the margins of Ethiopian society and shared in a quest for a homeland based on common ethnicity. Not surprisingly, both groups became involved in insurgencies against the state, which only further degraded the security situation. The Mengistu government's implementation of Marxist polices and move to the extreme left did little ease the tensions among ethnic groups. Mengistu steadfastly refused to compromise with these groups or cede effective control of "*one inch of national territory*," which was his stated policy (Westad 2007 p 259). As a result, especially in the Ogaden, repression of non–ethnic Ethiopians was stepped up, and the Somalis clearly took notice.

5 Discussion

This case study suggests that one of the most important prerequisites for a state to adapt to extreme environmental stress and climate disasters is political stability and sound governance (Westing 1991). Unfortunately, the outlook in this important area, especially in the developing world, is not encouraging as many of the world's developing countries are failing (Galgano 2007). In fact, an examination of contemporary history suggests that three quarters of all wars since 1945 have been within developing states and those violent conflicts have caused some 20 million deaths (Solow 2011). Certainly, environmental stress is not the sole cause of violent conflict, and without question, Somalia was poised to invade the Ogaden regardless of the environmental conditions in Ethiopia. Nevertheless, it is clear that Ethiopia's internal, political, and civil problems were pushed beyond the brink by a severe drought and devastating famine. The inability of the state to adapt prompted a mass migration into a sensitive area, which likely provoked the Somalis to attack in 1977 (Myers 2004). Therefore, it is not irrational to assume that Ethiopia's severe environmental problems were a fundamental and perhaps decisive factor in triggering this war. Nevertheless, there are simply too many variables, as this case study demonstrates, to develop a simple cause–and–effect model. However, failed states in the developing world certainly appear to be predisposed to environmentally triggered instability given their ineffective governments, weak economies, and social stratification (Smith and Vivekananda 2007). These factors combine to make them more susceptible to violence and instability enabled by an extreme environmental disaster such as the Ethiopian drought and famine (Westing 1991). In this case, the common Ethiopian peasant lived under great pressure from poor economic circumstances, increasingly hostile environmental conditions, and civil unrest (Ofcansky and Berry 1991).

Today there are about 2.7 billion in the developing world who live in absolute poverty and with the absence of effective governmental institutions (Smith and Vivekananda 2009). Projections indicate that environmental stress will have a fundamental effect on conflict because the economic welfare of more than 3.5 billion people—about one–half of the world's population—is tied to the land (Solow 2011). Therefore, factors such as agricultural productivity, water, fuel, fisheries, and forested lands are crucial environmental indicators; especially given the reality of population growth and climate change (U.N. 1987; IPCC 2012). This is of great consequence because nearly 75% of the world's most impoverished inhabitants are subsistence farmers attempting to live on land with declining productivity (IPCC 2014). As this case study suggests, drought, deforestation, and soil erosion can quickly become major problems in these regions, which are magnified by social iniquities and ineffective governments.

Understanding the complex linkages between environmental stress, disasters and conflict is essential for leaders in governmental and nongovernmental relief agencies and in their security forces. While a framework such as the one used in this case study cannot be predictive with any certainty, it can illustrate the complex interactions of environment, society, and government in a potential relief operation or military intervention. During the last decade, disasters have caused damage estimated at about $67 billion per year. Furthermore, the economic costs attendant to natural disasters have increased by a factor of 14 since the 1950s and these costs only promise to accelerate (IPCC 2012). With recent conflicts and environmental disasters in Indonesia, Rwanda, East Timor, Darfur, and Haiti as the precedent, the use of Western and United Nations (U.N.) relief and military forces to address the environmental dimensions of regional conflict has been now well established although U.N. and Western leadership has approached these commitments with acute reluctance—post earthquake response in Haiti is the most recent example (Dulian 2004; Wahlstrom 2013).

6 Summary and Conclusions

While the role of the environment as an enabler of violent conflict, on an interstate scale, remains a hypothetical exercise, environmental factors will continue to be strategically important variables on the national security landscape. Without question, the environment should be used as an indicator of impending regional instability and a persistent reminder of the significance of these variables to military and security affairs. This case study suggests that a reasonable nexus exists between environmentally induced instability and violent conflict—in certain scenarios. Undoubtedly, the Somalis maintained irredentist designs on the Ogaden region and their invasion may have been inevitable. However, as this analysis suggests, their decision to invade was possibly induced by what they saw as a unique window of opportunity represented by the devolution of the Ethiopian state. The combination of drought and famine, internal political conflict, and growing insurgencies clearly weakened Ethiopia. Finally, it is apparent that the Somalis misjudged the environmentally–forced migration of some three million Ethiopians into the Ogaden as an encroachment on their sovereign territory and a clear provocation.

The evolution of the global strategic situation following the end of the Cold War has led to the acceptance of an expanded definition of national security, which includes conflict intensified by environmental factors, population pressure, and resource scarcity. Globalization has contributed to environmental security problems because it has created expectations in the developing world of economic growth, which when combined with population pressure and climate change has stressed many ecosystems beyond carrying capacity. It has also raised the number of failed states incapable of keeping pace with exposure to environmental change. As global population grows, economic demands may exceed the resource and economic base of many states, erode governmental legitimacy, destabilize regions, and foster conflict over increasingly scarce resources.

The scenario presented in this paper suggests that the future is not bright. People already consume 40% of the world's food and energy potential. While that percentage may be sustainable, it is unlikely that it can keep pace with expected increases in global population and the economic and demographic changes attendant to globalization. Fortuitously, however, this grim prognosis is only a forecast. Like all predictions, it is rooted in contemporary trends and those of the recent past. Human society is not necessarily destined to enter a slow and painful decline into environmental chaos: there are scientific, technical, and economical solutions that may reduce the level of environmental stress and diminish potential conflict. Yet, possibilities for adaptation are confounded by substantial social, political, and institutional barriers, especially in the developing world. Unquestionably, our response to humanitarian disasters such as the one in the Ogaden demands the evaluation of environmental threats and using an analytical framework can be helpful. The principle challenge to U.S. governmental agencies and its security apparatus is that they are normally in the business of reacting to crises, and not traditionally staffed to plan for the non–traditional challenges presented by environmental security threats such as the one given in this case study. However, in order to lower the threshold of these environmental/humanitarian crises, the U.S. must address the roots of environmental instability. Thus, leaders within governmental and nongovernmental organizations must develop plans to deal with environmental security threats. This may mean adopting "softer" or stability–enhancing strategies that promote regional plans, rather than state–centric solutions, and assist in the development of governance and infrastructure in developing regions.

Literarture Cited

Alexander D (1993) Natural disasters: a framework for research and teaching. Disasters 15(3):209–226

Burke M, Miguel E, Satyanath S, Dykema J, Lobell D (2009) Warming increases the risk of war in Africa. Proc Natl Acad Sci 106(49):508–537

Christensen ER (ed) (2008) The Johns Hopkins and red cross–red crescent public health guide in emergencies. International Federation of Red Cross and Red Crescent Societies, Geneva. http://www.jhsph.edu/research/centers-and-institutes/center-for-refugee-and-disaster-response/publications_tools/publications/_CRDR_ICRC_Public_Health_Guide_Book/Public_Health_Guide_for_Emergencies. Accessed 11 July 2017

Conway D, Mould C, Bewket W (2004) Over one century of rainfall and temperature observations in Addis Ababa, Ethiopia. Int J Climatol 24:77–91

Cutter SL (1993) Living with risk. Edward Arnold, London

Cutter S, Golledge R, Graf WL (2002) The big questions in geography. Prof Geogr 54(3):305–317

Diehl P, Gleditsch N (eds) (2001) Environment and conflict. Westview Press, Boulder

Dulian A (2004) Rwandan genocide. Int Aff 50(4):40–44

Floyd R (2014) Analyst, theory, and security: a new framework for understanding environmental security. In: Floyd R, Matthew RA (eds) Environmental security approaches and issues. Routledge, New York

Galgano FA (2007) A geographic analysis of ungoverned spaces. Pennsylvania Geogr 44(2):67–90

Galgano FA, Krakowka A (2010) The environment's influence on regional stability and conflict. Pennsylvania Geogr 48(2):11–40

Guha–Sapir D, Hargitt D, Hoyois P (2004) Thirty years of natural disasters 1974–2003: the numbers. Presses University of Louvain, Louvain–la–Neuve

Harkavy RE, Neuman SG (2001) Warfare in the third world. Palgrave, New York

Hendrix C, Salehyan I (2012) Climate change, rainfall, and social conflict in Africa. J Peace Res 49(1):35–50

Homer–Dixon T (1991) On the threshold: environmental change as causes of acute conflict. Int Secur 16(2):76–116

Homer–Dixon T (1999) Environmental scarcity and violence. Princeton University Press, Princeton

Hsiang S, Meng K, Cane M (2011) Civil conflicts are associated with global climate change. Nature 476:438–441

Intergovernmental Panel on Climate Change (IPCC) (2012) Managing risks of extreme events and disasters to advance climate change adaptation. Special report of working groups I and II of the Intergovernmental Panel on Climate Change, Field CB, Barros V, Stocker TF, Qin D, Dokken DJ, Ebi KL, Mastrandrea MD, Mach KJ, Platter GK, Allen SK, Tignor M, and Midgley PM (eds), Cambridge University Press, Cambridge. http://ipcc-wg2.gov/SREX/. Accessed 21 July 2017

Intergovernmental Panel on Climate Change (IPCC) (2014) Climate change 2014: impacts, adaptation, and vulnerability. Summary for policy makers. Working Group II Contribution to the Fifth Assessment Report of the Intergovernmental Panel on Climate Change. Cambridge University Press, Cambridge. http://www.ipcc.ch/report/ar5/wg2/. Accessed 8 Nov 2016
Kalif MH (2000) The politics of famine in the Ogaden. Rev Afr Polit Econ 27(84):333–337
Kaplan RD (2000) The coming anarchy: shattering the dreams of the post-cold war. Random House, New York
Klare M (2002) Resource wars: the new landscape of global conflict. Henry Holt and Company, New York
Krakowka A, Hiemel N, Galgano FA (2012) Modeling environmental security in Sub-Saharan Africa. Geogr Bull 53:21–38
Lee J (1997) An inventory of conflict and environment, a conceptual model. School of International Service, American University. http://www1.american.edu/ted/ice/elements.htm. Accessed 1 June 2016
Mathews J (1989) Redefining security. Foreign Aff 68(2):162–177
Myers N (2004) environmental security: What's new and different? The Hague conference on environment, security and sustainability, the peace palace, The Hague, The Netherlands, 9–12 May 2004. http://www.envirosecurity.org/conference/working.php. Accessed 18 June 2017
NCDC (National Climate Data Center) (2013) Monthly climatological summary for Ethiopia, 1964–1987. Climate Data Online, U.S. Department of Commerce, National Oceanographic and Atmospheric Administration. http://www.ncdc.noaa.gov/cdo-web/quickdata. Accessed 19 Mar 2013
Nkaisserry J (1997) The Ogaden war: analysis of its causes and its impact on regional peace on the Horn of Africa. MA Thesis, United States Army War College, Carlisle
Ofcansky T, Berry L (1991) Ethiopia: a country study. Kessinger Publishing, New York
Porter G (1995) Environmental security as a national security issue. Curr Hist 94(592):218–222
Reuveny A (2007) Climate change-induced migration and violent conflict. Polit Geogr 26(6):656–673
Schwab P (1978) Cold war and the Horn of Africa. Afr Aff 77(306):6–20
Smith D, Vivekananda J (2007) A climate of conflict. International Alert, London. http://www.international-alert.org. Accessed 12 Dec 2017
Smith D, Vivekananda J (2009) Climate change, conflict and fragility. International Alert, London. http://www.alert.org/resources/publications/climate-change-conflict-and-fragility. Accessed 1 Aug 2017
Solow A (2011) Climate for conflict. Nature 476:406–407
Tareke G (2000) The Ethiopia–Somalia war of 1977 revisited. Int J Hist Stud 33(3):635–667
Trondalen JM (2009) Climate changes, water security, and possible remedies for the Middle East. United Nations World Water Assessment Programme, Scientific Paper, United Nations Educational, Scientific and Cultural Organization, Paris
U.N. (United Nations) (1987) Our common future. Report of the world commission on environment and development (The Brundtland Report). Oxford University Press, New York. http://www.un.org/documents/ga/res/42/ares42-187.htm/. Accessed 23 June 2012
Wahlstrom M (2013) Progress and challenges in global disaster reduction. Int J Disaster Risk Sci 4(1):48–50
Westad OA (2007) The global cold war. Cambridge University Press, Cambridge
Westing A (1991) Environmental security and its relation to Ethiopia and Sudan. Ambio 20(5):168–171

The 1994 Rwandan Genocide

Amy K. Richmond and Francis A. Galgano

Abstract Links between human factors, governance, and environmental degradation have the potential to trigger internecine conflict, such as the 1994 genocide in Rwanda. Evidence suggests that this trend will persist in the near term because climate change and the adverse effects of the environment will continue to stress marginal environments in places with inherently weak governance. This has led to a greater acceptance of the environment as an emerging factor on the national security landscape. This analysis uses a framework that encompasses natural and anthropogenic factors to enable a comprehensive analysis of issues that contribute to environmentally triggered conflict. In the case of the 1994 Rwandan civil war, factors of population, economic decline, drought, and unstainable agricultural practices exacerbated latent ethic conflict to spark one of the worst human disasters of the last century.

Keywords Adaptation · Climate change · Drought · Ecocide · Environmental security · Ethnic warfare · Genocide · Governance · Rwanda · Sustainability · World Bank

1 Introduction

The relationship between violent conflict and environmental degradation is a matter of some polemic. Events in recent history have demonstrated a relationship between environmental stress and conflict because in combination they intensify economic, ethnic, societal, and political fissures, thus undermining security. Evidence suggests that this trend will persist because climate change and the adverse effects of the

A. K. Richmond (✉)
Department of Geography and Environmental Engineering, United States Military Academy, West Point, NY, USA
e-mail: amy.richmond@usma.edu

F. A. Galgano
Department of Geography and the Environment, Villanova University, Villanova, PA, USA
e-mail: francis.galgano@villanova.edu

F. Galgano (ed.), *The Environment-Conflict Nexus*, Advances in Military Geosciences, https://doi.org/10.1007/978-3-319-90975-2_10

environment will continue to stress marginal environments in places with inherently weak governance (IPCC 2014). This has inspired an acceptance of the relationship between regional instability and environmental factors, and has led to a greater acceptance of the environment as an emerging factor on the national security landscape—the so–called environmental security paradigm (Butts 1994; Homer-Dixon and Levy 1995).

The Cold War security landscape usually meant that adequate control was exerted over proxy regimes to preclude regional hot spots from escalating into violent confrontations. The recent proliferation of ungoverned space, however, combined with diluted superpower control over former surrogate regimes has enabled festering ethnic and political enmities to erupt into violent conflicts often triggered by environmental stress (Renner 2002). This problem is aggravated by unsustainable environmental practices, population pressures, migration, and resource shortages (Gleditsch et al. 2007). With conflicts in Somalia, Rwanda, East Timor, Haiti, and now Syria as the precedent, the use of Western military forces to address humanitarian dimensions of regional conflict has been now well established (Dulian 2004; Femina and Werrell 2012). Nevertheless, conflicts with an environmental component coupled with divisive ethnic dimensions, such as those observed in 1994 Rwanda, have increased pressure on the West to commit military forces and other resources to stability efforts (Drapeau and Mignone 2007). In fact, strategic policy documents produced by the U.S. National Security Council (NSC 1991) and U.S. Department of Defense (DoD 2005) have delineated a long–standing U.S. strategic interests in environmentally enabled instability. The environment became a clear element in U.S. strategy in 1991, when the NSC pointed out that, "*stress from environmental challenges is already contributing to political conflict*" (NSC 1991, 2). By 2005, the DoD identified environmentally related instability as a fundamental strategic concern because of evidence that environmental stress is an important contributor to contemporary conflicts. Furthermore, environmental conflict typically manifests itself along ethnic lines, thus making its international management difficult (DoD 2005).

The interplay among factors of climate change, environmental degradation, resources shortages, and politico–military dynamics are highly problematic in much of the world, but especially in failed and failing states. Thus, it is plausible that we will witness a surge in three modes of conflict: internecine conflict driven by resources shortages, environmental stress, and demographic trends; civil war prompted by governmental collapse and economic factors; and limited–scale interstate conflicts. This chapter presents a case study that examines the initiation of a violent internecine conflict between tribal groups, and ultimately a genocide in Rwanda in 1994. Rwanda was once thought to be the architype of the new African state. It was thought to have a steady and prosperous economy and a stable society. However, all of that changed in 1994 when the country devolved into violent civil warfare and a genocide that ultimately witnessed the death of some one million people. Although the conflict was triggered by long–standing latent ethnic hatreds and economic collapse, environmental factors clearly exacerbated the situation and contributed to instability and violent civil conflict (Homer-Dixon 1999).

2 The Environment and Conflict

The incidence of environmentally triggered conflict is not new. Researchers have offered compelling evidence that identifies the role of environmental stress in precipitating warfare in ancient China as well as the collapse of the Mayan, Anasazi, and Akkadian civilizations (Gibbons 1993; Abate 1994; Diamond 2005; Mays 2007; Zhang et al. 2007). Recent examples in Darfur and Syria indicate the nexus of environmental stress and violent conflict is a reality and that the specter of contemporary environmental change and resource scarcity may prompt an increase in violent conflict (Femina and Werrell 2012). However, environmental stress alone does not trigger violent conflict, nor is an exclusively causative factor. Evidence suggests that it exacerbates conditions when it is combined with weak governance and social fragmentation, to affect an escalation of violence, typically along latent ethnic divisions. Furthermore, the spatial distribution of contemporary environmental stress is pervasive, but not uniform. Contemporary trends indicate that environmentally driven violence is concentrated in the developing world or regions with extreme social fragmentation (Homer-Dixon 1999; Gleditsch et al. 2007). Developing states are more susceptible to environmentally triggered violent conflict because they are, characteristically, more dependent on the environment for their economic productivity have manifested persistently weak governance (Homer-Dixon 1999; Galgano 2007).

Weak governance is a seminal problem in the developing world, and since 1990, the number of failing states has increased. The World Bank (Kaufmann et al. 2003) examined effective sovereignty and governance by monitoring six key metrics as a means of quantifying the level of effective state control. Their findings suggest that of 187 states examined, 92 exhibited considerable levels of political instability and can be categorized as failing states. In the category of *government effectiveness*, 75 states exhibit significant levels of failure, and 20 of those showed an alarming drop in government control (Kaufmann et al. 2003). This trend has continued during the past decade and evidence now suggests that nearly 60% of the states in the developing world experience critical governance issues and are unable to adapt to climate change or other environmental shocks (FSI 2017). Ungoverned space is problematical because these places have large areas that are outside of effective government control and thus, can be affected severely by humanitarian disasters, environmental stress, and ethnic conflict. This is because they typically lack effective institutions and the financial and material resources to safeguard the population from the effects of environmental stress (Galgano 2007). This raises the complexity of the problem for government leaders as well as directors of non-governmental organizations and intergovernmental bodies as they attempt to develop relief strategies, especially without an effective framework for understanding the nature of the conflict and its environmental underpinnings (deMenocal 2001).

Environmental stress will play a pervasive role in future conflicts because the economic well-being of about one–half of the world's population is tied directly to the land, thus making agricultural space, water, fuel, and forested space critical environmental indicators, especially considering anticipated population growth and projected climate change (IPCC 2014). This is important because nearly two billion people do not have access to clean drinking water (Gleick 2008), and nearly 75% of the world's most impoverished inhabitants are subsistence farmers. Drought, desertification, deforestation, soil erosion, and exhaustion are major problems in these regions. These are compounded by the fact that although agricultural space, biomass fuels, and water are renewable resources, in many places non–sustainable practices are depleting them far beyond their renewal capacity (Homer-Dixon 1999). This scenario is made more problematical because anticipated population growth, especially in the developing world, will result in higher per capita consumption rates, exacerbating extant non-sustainable practices.

3 Framework for Analysis

Since 1990, violent conflicts have occurred in Bosnia, Kosovo, Croatia, Rwanda, Ivory Coast, Burundi, Angola, Nigeria, Sudan, Turkey, Azerbaijan, Georgia, Kashmir, Myanmar, Sri Lanka, Iraq, Palestinian Territories, Darfur, Libya, and Syria (Kaufman 1996; Renner 2002; Femina and Werrell 2012). It would be mistaken to assert that environmental stress instigated these conflicts, and too difficult to disaggregate their human and environmental components because they are interrelated and complex. Consequently, if we attempt a proactive approach to mitigate environmentally enabled conflict, it is useful to employ an analytical framework from which we can make informed assessments (Butts 1994). This framework must account for anthropogenic and natural environmental processes and recognize fundamental ethnic, economic, cultural, and political issues behind regional instability, understanding that each place is different (Homer-Dixon 1999).

Detractors of this perspective of conflict analysis argue that they result solely from politico-military factors, which are minimally influenced by environmental stress, and allude to environmental determinism—this too is an unsophisticated approach. Environmental stress and scarcity result from the combined influence of anthropogenic effects on the environment in conjunction with the vulnerability of the ecosystem. Scarcity and stress contribute to conflict only under certain circumstances, but there is no deterministic link (Percival and Homer-Dixon 1995). Not all violent conflicts are alike and the influence of environmental stress on warfare will vary in magnitude from example to example. Nonetheless, there is compelling evidence that indicates that environmental stress does enable violent conflicts (Gibbons 1993; Abate 1994; Homer-Dixon 1999; Renner 2002; Diamond 2005; Gleditsch et al. 2007; Mays 2007; Zhang et al. 2007; IPCC 2014).

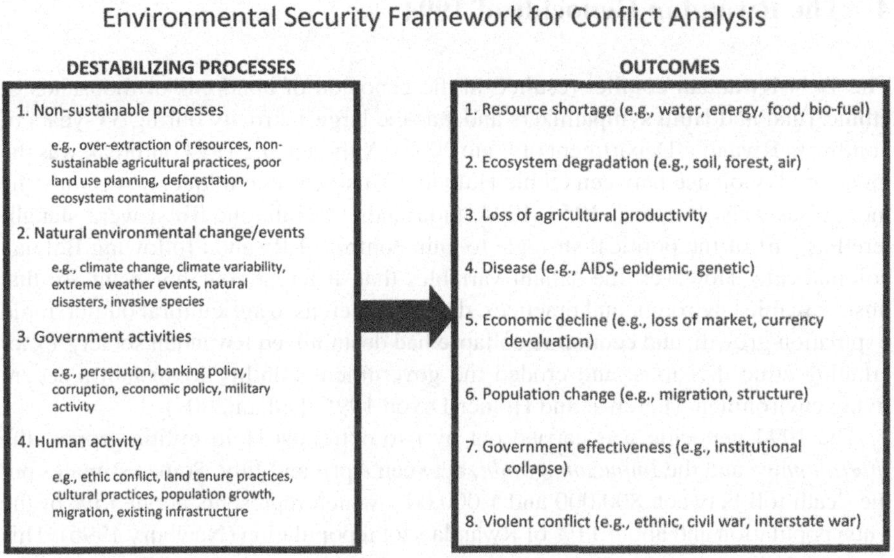

Fig. 1 The environmental security framework for conflict analysis

The relationships between food, population growth, migration, climate, environmental resources, and environmental stress are evident in many developing states (Homer-Dixon 1999). This Malthusian paradigm generates much disagreement among scholars. Nevertheless, both sides have to agree on an irrefutable outcome that was evident in Rwanda in 1994—population growth and environmental stress, superimposed over latent ethic and political divisions will, in the end, be solved one-way or another. Thus, we use a framework for conflict analysis that identifies sources of environmental stress and linkages to political, cultural, economic, and ethnic dimensions. The framework is developed after the work of Butts (1994), Percival and Homer-Dixon (1995), Homer-Dixon (1999)), and Diamond (2005). The framework is not intended to be predictive or proscriptive, only an analytical tool to account for dynamic and complex factors that contribute to environmentally enabled conflict.

The framework (Fig. 1) is not a checklist because such inventories tend to disconnect intrinsically linked factors into artificially discrete variables. Instead, it forms the basis for a narrative explaining links between environmental stress and conflict. The framework suggests that environmentally triggered conflict evolves from four fundamental processes: (1) unsustainable practices; (2) natural environmental change; (3) governance; and (4) human activity. These processes form the basis of the framework and are not mutually exclusive: they take place concurrently, and occur at the intersection of the natural landscape and human activity. So too are the outcomes: (1) resource shortages; (2) ecosystem degradation; (3) loss of agricultural productivity; (4) disease; (5) economic decline; (6) population change; (7) decline of effective governance; and (8) violent conflict.

4 The Rwandan Genocide of 1994

The 1994 Rwandan conflict resulted in the genocide of hundreds of thousands of ethnic Tutsi and Hutu sympathizers and was the largest atrocity during 35–years of conflict in Rwanda (Department of State 2002). Although the 1994 outbreak was the most vivid, violence between ethnic Hutu and Tutsi was not unique; it was only the most disastrous. Between 1959–1994, thousands of Hutu and Tutsi were slaughtered as part of the political struggle to gain control of Rwanda following Belgian colonial rule. However, the salient variables that differentiates this event are that unsustainable environmental practices, drought, decreased agricultural output, rapid population growth, and economic collapse had destabilized Rwandan society, exacerbating ethic divisions, and eroded the government's ability to sustain a secure living environment (Percival and Homer-Dixon 1995; Dulian 2004).

The 1994 genocide was carried out by two extremist Hutu militia groups, the *Interahamwe* and the *Impuzamugambi*, between April and July. Some estimates put the death toll between 800,000 and 1,000,000, which represents about 75% of the Tutsi population and about 11% of Rwanda's total population (Newbury 1995). This scale of killing is particularly gruesome because the weapon of choice was the machete, and the U.N. estimates that 1000 Rwandans were hacked to death every 20 min during the height of the genocide (Melvern 1999). So then, how did the nexus of ethnic friction, ineffective governance, and environmental stress come together to enable mass-murder?

4.1 *Society and Governance*

The distinction between Hutu and Tutsi is not expressly ethnic or racial, but rather economic. The Hutu are farmers and the Tutsi pastoralists who were more economically prosperous. Tutsi dominated the region politically and simply labeled the indigenous people "Hutu." Hence, Hutu came to be a trans-ethnic identity associated with subjugation (Mitchell 1997). The Belgians, seeking an explanation for the complex society they found in the colony, framed the Hutu–Tutsi distinction as one of race. They issued racial identification cards, giving preferential treatment to Tutsis for administrative and economic positions, further deepening this latent ethnic division (Newbury 1995). When independence came in 1961, this pseudo–racial divide propelled Rwanda into a 35–year period of civil unrest, conflict, and forced migration, punctuated by periods of relative calm (Diamond 2005).

The downing of the Rwandan President's plane on 6 April 1994 served as the political catalyst for the start of the genocide, and the killing was well organized with government support. By the time it started, Hutu militia was 30,000 strong and organized nationwide. Furthermore, there were widespread examples of government soldiers participating in the violence as well (Department of State 2002). The orgy of violence was to continue—as the West stood by—until the Rwandan

Patriotic Front (RPF) seized control of the country and effectively ended the violence in mid-July 1994 (Mitchell 1997).

4.2 Population Growth

Perhaps the seminal environmental stressor in Rwanda was its population density and the spiral of ecosystem degradation it engendered. Rwanda is situated in East Africa (Fig. 2), which has the highest regional population growth in the world, averaging about 4.1%, with a doubling time of 17 years. Rwanda is one of the world's most densely populated states, and the most densely populated in Africa: its 1995 population was about 7.5 million with a growth rate of 4.7%. To make matters worse, large internal refugee displacements pushed increasing numbers of people into environmentally sensitive areas (Percival and Homer-Dixon 1995).

Explosive population growth was enabled by a number of interrelated human factors to include adoption of non–native crops, improved sanitation and health care, and relative political stability, which opened up once contested areas for farming. Characteristics of Rwanda's natural landscape facilitated population growth as well. Rwanda has comparatively higher relief and moderate temperatures. This is important because lower temperatures curtail vector populations, thus limiting the spread of disease (Diamond 2005). Rwanda also experiences moderate amounts of rainfall in two seasonal peaks (Fig. 2), which permits year-round farming, and limits nutrient leaching. Furthermore, its volcanic soils are fertile, and the geologic structure establishes conditions for a sustainable groundwater supply. Therefore, sustainable farming practices should enable Rwandans to produce crops to support a growing population (Mitchell 1997; Diamond 2005).

By 1994, however, population growth, migration, and unsustainable farming practices dangerously degraded the environment and diminished food output. Rwanda's population growth meant that there was little useful arable land left for cultivation. Furthermore, about half of all farming in Rwanda is conducted on hill slopes greater than 10–degrees. Over–cultivation and cultivation on marginal land led to increased erosion and decreased soil fertility (Clay et al. 1998). Thus, Rwanda was transformed from one of the region's leading producers in *per capita* food production in 1980, to one of its worst by 1990. Although total food output increased by 10% between 1980 and 1990, *per capita* production declined nearly 20% (Percival and Homer-Dixon 1995). The basic problem was that population growth was nearly exponential, but its food production increased at something approximating a linear rate that had begun to decline. Environmental stress damaged the agricultural system and food production decreased. Consequently, as food became scarce and internal migration became commonplace, a dangerous strain was placed on latent ethic divisions (Diamond 2005).

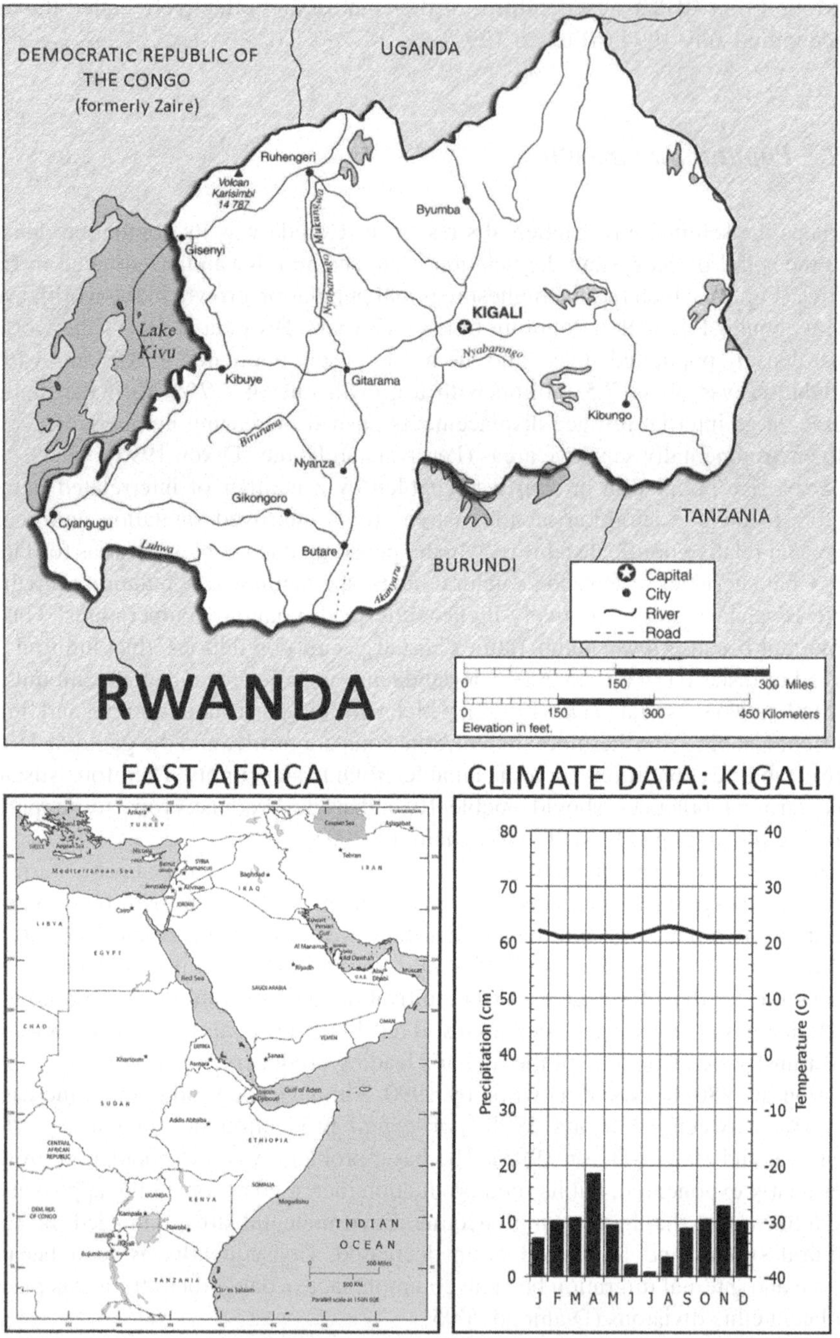

Fig. 2 Map of Rwanda and East Africa

4.3 Ecocide

The foundations of Rwanda's environmental degradation are associated with rapid population growth and a series of unsustainable land use practices, which were underscored by a decade–long drought. By 1990, Rwanda had experienced 15 years of relative calm and prosperity. It established a trade surplus built on coffee and tea exports, and attracted substantial development investment by the World Bank. By every measure, it appeared that Rwanda was a model African state. Events conspired, however, to destroy this illusion and trigger the genocide. The first was a steep decline in world coffee and tea prices, which the Rwandan economy could not absorb, and subsequently collapsed, which was followed by withdrawal of external monetary support. The second factor was environmental, which quickly exposed extant unsustainable land use practices that over–stressed the environment. During the early 1990s, a major drought affected East Africa and Rwanda in particular. Rainfall totals declined by as much as 30%, but more importantly, the drought revealed a set of fundamental land use problems, namely deforestation, soil erosion, and soil exhaustion (Newbury 1995; Diamond 2005).

Rwanda's large population depended on subsistence farming, which was pushed into marginal land. Remarkably, even though Rwanda was a relatively prosperous state; population growth was not offset by new farming technology and more efficient practices. Instead, more land was placed under the plow; and thus, by 1985, all arable land in Rwanda, discounting that devoted to parks and other government land use, was under cultivation (Percival and Homer-Dixon 1995). To make this happen, forests were clear–cut and marshes were drained, exposing slopes to runoff and erosion, thus limiting the percolation of precipitation into the ground water table. Fallow periods were shortened resulting in soil exhaustion. Deforestation led to severe soil erosion and a lowering of the water table to the extent that streams began to run dry. Hence, when the climate began to shift into a decadal drought cycle, the ability to irrigate was lost. Clear–cutting was also the result of Rwanda's high fuel–wood consumption, which was using 2.3 million cubic meters more than it was producing (Mitchell 1997).

The final variable in this ecocide was land tenure practices. The normal farm plot in 1980 was about 2.5 acres per family. As the population grew, additional arable land was tilled, yet food production could not accommodate the number of people to feed. Furthermore, the average farm plot was reduced to less than one acre per family as custom normally dictated that the oldest son would inherent the family farm. Once all of the arable land was taken, younger siblings had to remain at home; hence, family farms were split almost infinitesimally, and larger families had to be fed by increasingly small and unproductive farm plots (Clay et al. 1998).

The basic components of Rwanda's environmental security framework are summarized in Fig. 3. The seminal problem was that Rwanda had a large, densely concentrated population that depended on the land for resources that were rapidly diminishing resulting from non-sustainable practices and environmental change. About 90% of Rwandans were engaged in subsistence agriculture on increasingly

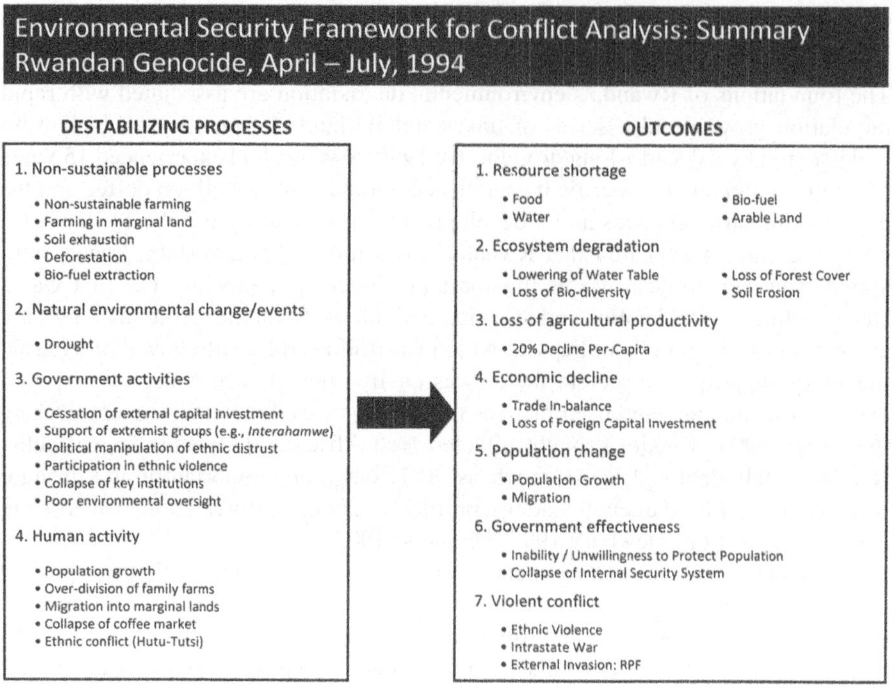

Fig. 3 Framework summary for Rwanda

smaller plots of land with declining productivity and by 1994, population had clearly outpaced food production. These environmental factors, combined with the collapse of the global coffee market, as well as latent ethic problems and political competition perhaps pushed this society over the edge.

5 Summary and Conclusions

Environmental security is, for the purpose of this analysis, defined as a process involving, environmental risk analysis based on an understanding of the complex interactions between anthropogenic and natural processes that destabilize the environment and contribute to instability or conflict. To operationalize this definition we use an analytical framework that is not proscriptive or predictive, rather it is a method for organizing the varied, dynamic, and complex environmental factors that produce regional instability and enable violent intrastate conflict. The framework for environmental security analysis is intended to form the foundation for a narrative explaining links between environmental stress and conflict to include: non–sustainable practices that degrade ecosystems; natural environmental change or events; governmental practices; and human activity.

No two conflicts are the same, thus accentuating the need for careful assessments of regional and local conditions to understand how ethnic, economic, and political tensions are affected by environmental stress. The Rwanda example demonstrates the fundamental problem with this type of analysis: factors of the human landscape and environmental stress are not mutually exclusive. They are complicated and inextricably linked making it difficult to quantify how much the environment has enabled a violent conflict. Nonetheless, the Rwanda example also demonstrates that population growth and severe environmental stress caused by non-sustainable practices, superimposed over latent ethic and political divisions will, in the end, be solved sometimes very violently.

Complex, interacting factors enabled Rwanda's genocide, and military geography offers compelling insight and an especially valuable vantage point from which to conduct an analysis of conflict and environmental security. Rwanda's genocide, like other violent conflicts, had profound roots in long-standing ethnic distrust, politically charged manipulation, and weak governance; but, economic disparity and environmental stress were certainly contributing factors. In Rwanda's case, a principally rural society that depended heavily on subsistence agriculture and resource extraction from the environment proved particularly vulnerable to the effects of environmental stress, which threatened its stability even before the 1994 genocide. Evidence suggests, however, that by 1994, these stresses, combined with economic collapse and weak governance, precipitated a three-month-long episode of violence the likes of which the world has seldom witnessed.

References

Abate T (1994) Climate and the collapse of civilization. Bio Science 44(8):516–519

Butts KH (1994) Environmental security: a DoD partnership for peace, Strategic Studies Institute Special Report, U.S. Army War College Press, Carlisle

Clay D, Reardon T, Kangasniemi J (1998) Sustainable intensification in the highland tropics: Rwandan farmer's investment in land conservation and soil fertility. Econ Dev Cult Chang 46(2):351–377

deMenocal PB (2001) Cultural responses to climate change. Science 292(5517):667–673

Department of Defense (DoD) (2005) The national defense strategy of The United States of America. U.S. Government Printing Office, Washington, DC

Department of State (2002) Department of State legal analysis of 1994 Rwandan genocide. Am Journof Intnl Law 96(1):258–262

Diamond J (2005) Collapse. Penguin Books, New York

Drapeau MD, Mignone BK (2007) Culture, conflict, and … climate? Science 316(5831):1564

Dulian A (2004) Rwandan genocide. Int Aff 50(4):40–44

Femina F, Werrell CE (2012) Update: climate and security 101: chy the U.S. national security establishment takes climate change seriously. The Center for Climate and Security, Briefer no. 23, 26 February 2014. https://climateandsecurity.org/reports/#BRIEFERS. Accessed 12 Jan 2017

Fragile States Index (FSI) (2017) Fragile states index 2017: factionalization and group grievance fuel rise in instability. http://fundforpeace.org/fsi/category/analysis/fsi17/. Accessed 21 Jan 2018

Galgano FA (2007) A geographic analysis of ungoverned spaces. Pennsylvania Geogr 44(2):67–90
Gibbons A (1993) How the Akkadian empire was hung out to dry. Science, News Services, 261(5124):985
Gleditsch NP, Nordås R, Salehyan I (2007) Climate change and conflict: the migration link. Coping with Crisis Working Paper Series. International Peace Academy. http://www.ipacademy.org. Accessed 31 Aug 2008
Gleick P (2008) The world's water, 2006–2007. Island Press, Chicago
Homer-Dixon TF (1999) Environmental scarcity and violence. Princeton University Press, Princeton
Homer-Dixon TF, Levy MA (1995) Environment and security. Int Secur 20(3):189–198
Intergovernmental Panel on Climate Change (IPCC) (2014) Climate change 2014: impacts, adaptation, and vulnerability. Summary for policy makers. Working Group II Contribution to the fifth assessment report of the Intergovernmental Panel on Climate Change. Cambridge University Press, Cambridge. http://www.ipcc.ch/report/ar5/wg2/. Accessed 8 Nov 2016
Kaufman C (1996) Possible and impossible solutions to ethnic civil wars. Int Secur 20(4):136–175
Kaufmann D, Kraay A, Mastruzzi M (2003) Governance matters III: Governance indicators for 1996–2002. The World Bank, World Bank Institute, Global Governance Department and Development Research Group
Mays LW (2007) Water sustainability of ancient civilizations in Mesoamerica and the American Southwest. Water Supply 7(1):229–236
Melvern L (1999) The record of killing by machete was 1,000 every 20 minutes. The Times (London) 4 March:15
Mitchell T (1997) Rwanda and conflict. International Conflict and the Environment: Rwanda Case, Case No. 23, Trade and Environment Database. http://www.american.edu/ted/ice/Rwanda.htm. Accessed: 3 Sept 2008
National Security Council (NSC) (1991) National security strategy of the United States. US Government Printing Office, Washington, DC
Newbury C (1995) Background to genocide: Rwanda. Issue: A J Opin 23(2):12–17
Percival V, Homer-Dixon T (1995) Environmental scarcity and violent conflict: The case of Rwanda. Occasional Paper. Project on Environment, Population and Security, Washington, DC.: American Association for the Advancement of Science and University of Toronto. http://www.library.utoronto.ca/pcs/eps/Rwanda/rwanda1.htm. Accessed: 28 July 2012
Renner M (2002) The anatomy of resource wars, Worldwatch Paper 162. In: Prough T (ed) State of the World Library. Worldwatch Institute, Danvers
Zhang DD, Zhang J, Lee HF, He Y (2007) Climate change and war frequency in Eastern China over the last millennium. Hum Ecol 35:403–414

Climate and the Syrian Civil War

Mark R. Read

Abstract The relationship between climate change and security represents a subset of environmental security, and a new area of multidisciplinary research. Although climate change is now discussed in the national security policy and doctrine of a number of countries, research linking climate change and security has been debated, and has not yet adequately dealt with high levels of uncertainty in the associated complex socio-environmental systems. This chapter examines the relationship between climate change and security using the Syrian civil war as a case study. Some policy makers, commentators, and scholars have proposed that the Syrian civil war was caused, at least in part, by a long–term regional drought during the late 2000s, and that the drought is attributable to anthropogenic climate change. Others critique such claims, arguing that the effects of the drought, especially drought–induced migration, have been exaggerated, that it is not possible to attribute the drought to climate change (natural or anthropogenic), or that it is simply inaccurate to attribute the civil war to the drought in light of other, more significant socio–political factors. This chapter explores claims on opposing sides of the issue, in order to illuminate the ongoing debate. Finally the chapter summarizes lessons from the Syrian civil war as an environmental security case study, and suggests areas for future research.

Keywords Syrian civil war · Environmental security · Climate change

1 Introduction

The Arab Spring began in December 2010 with street protests in Tunisia. During the ensuing months, protests spread to countries across North Africa and the Middle East. Of the many countries affected directly by the 2011 Arab Spring uprisings, Syria has faced the most devastating outcomes. To date, casualty estimates in the Syrian civil war include between 300,000 and 500,000 killed, and as many as 12 million—half of

M. R. Read (✉)
Department of Geography and Environmental Engineering, United States Military Academy, West Point, NY, USA
e-mail: mark.read@usma.edu

F. Galgano (ed.), *The Environment-Conflict Nexus*, Advances in Military Geosciences, https://doi.org/10.1007/978-3-319-90975-2_11

the country's pre-war population—have been displaced from their homes. Neighborhoods in many cities lay in rubble, the Syrian economy is in ruins, and the outcome of the conflict remains uncertain. Armed conflict spilled across Syria's eastern border with Iraq, when the Islamic State in Iraq and Syria (ISIS) launched a campaign from northeast Syria to re-establish territorial control over large areas of northern Iraq. Regional state actors, including Turkey, Lebanon, Israel, Jordan, Saudi Arabia, Iraq, Qatar, and Iran, have become involved in the conflict to varying degrees. Non–state actors, including Hezbollah and Kurdish groups, are also involved; and the international community, including the United States, Russia, and the United Nations, have committed people and resources to Syria. Numerous peace negotiations have stalled or failed. In 2017, Syria ranked the fifth most fragile country of 178 listed on the Fragile States Index, with economic, political, social, and cohesion indicators all continuing to worsen since 2011 (Fund for Peace 2017).

Many factors contributed to the outbreak of armed conflict in Syria: i.e., widespread and long–standing discontent with the Assad regime, sectarian friction between a Sunni majority and the ruling Shia (Alawite) minority, high unemployment, the success of Arab Spring movements in changing regimes in nearby states such as Tunisia and Egypt, and a response to a heavy–handed government crackdown on early Arab Spring protesters in Syria (Sorenson 2016). Another factor that may have contributed to the conflict, but that is not often discussed, is the long term (2006–2010) drought in agricultural regions of northeast Syria. Some have argued that the drought is attributable to climate change. Others downplay the social and political implications of the drought, arguing it should not be considered among the factors contributing to the Syrian civil war.

This chapter examines climate change, security, and conflict, using the Syrian civil war as a case study. The first section examines the relationships among climate change, security, and conflict, including perspectives from both the academic and policy communities. The second highlights the role that climate change and drought may have played in the Syrian civil war, discussing different perspectives on the ongoing debate. The chapter concludes with a summary of environmental security lessons gathered from the Syrian civil war, and recommends areas of future research.

2 Climate Change, Security, and Conflict

The relationship among climate change, security, and conflict is complex, as described in a relatively small body of scholarly research that has recently emerged. Once thought to be too slow-changing to serve as a security concern, climate change did not enter as a variable in the environmental security debate until the early– to mid–2000s. Barnett (2003) provides one of the first comprehensive discussions of climate change as a security issue, cautiously suggesting that framing climate change as a security issue (at least in part) may help bridge science and policy. A number of more recent studies have examined possible correlations between changes in climate and violent conflict in particular. Most of this research has grown out of the political science community, and employs traditional empirical methods (e.g., Nordas and Gleditsch 2007; Raleigh and Urdal 2007; Hsiang et al. 2013). Nonetheless, causal connections between climate change and conflict remain unclear, and are the subject of ongoing debate among scholars. In the end, most of the existing research attempting to link climate change with conflict does little to address the uncertainty surrounding future climate change and security, and has been generally insufficient to meet the demands of the policy community.

The Fourth Assessment Report (AR4) of the Intergovernmental Panel on Climate Change (Parry et al. 2007) briefly mentions links between climate change and conflict, but does not provide supporting research. The Fifth Assessment Report (AR5) includes a chapter on human security, and briefly addresses research on the relationship between climate change and armed conflict. The AR5 concludes that "*confident statements about the effects of future changes in climate on armed conflict are not possible given the absence of generally supported theories and evidence about causality*" (Adger and Pulhin 2014). Causal connections between climate change and conflict remain unclear, and are the subject of ongoing debate among scholars. Much of the scholarly research that has focused on climate and conflict has employed methods that attempt to correlate climate change with indicators of conflict. As noted in AR5, such methods are problematic—they typically focus on a single spatial level, thereby missing or neglecting emergent properties; they are limited in the number and scope of social variables included in their analysis; they normally follow linear causal relationships and fail to address non–linearity, feedback, or thresholds; and the existing research relies almost exclusively on data from past events and historic conditions, so the results may be very limited in what they can tell us about the future under conditions of unprecedented climate and social change (Adger and Pulhin 2014).

A growing body of non–scholarly literature addresses climate change, conflict, and security, including numerous white papers, studies, and other publications by government institutions, intergovernmental organizations (IGOs), and non–governmental organizations (NGOs). The U.S. Congress required the establishment of the U.S. Global Change Research Program in 1990, which coincided with some of the earliest research on the relationship between climate change and security. In 2008, Congress directed the Department of Defense to include discussion of the effects of climate change on U.S. national security in the next Quadrennial Defense Review (National Defense Authorization Act for Fiscal Year 2008). In 2017, Congress stated that climate change is "*a direct threat to the national security of the United States and is impacting stability in areas of the world both where the U.S. Armed Forces are operating today, and where strategic implications for future conflict exist*" (National Defense Authorization Act for Fiscal Year 2018). U.S. national strategic policy documents, including the 2010 and 2015 National Security Strategy, the 2010 and 2014 Quadrennial Defense Reviews, and the 2010 Quadrennial Development and Diplomacy Review, address the relationship between climate change and security. Each of these recurring strategic documents features climate change as an important factor influencing U.S. national security, but the theoretical underpinnings of climate change–security issues in the documents are unclear—the strategic documents neither list supporting references, nor contain any theoretical background or empirical evidence in the documents themselves. The policy community in the U.S. remains engaged on issues related to climate change and security. Within the executive branch, the national security community has conducted or funded a number of studies to better understand the national security implications of climate change. In 2009, the U.S. Navy created Task Force Climate Change to coordinate research and operational planning related to climate change impacts on maritime operations, especially in the Arctic.

Several NGOs in Washington, D.C., remain engaged on climate change–security issues, including the Center for a New American Security, CNA, the Center for Strategic and International Studies, the Brookings Institute, the American Security Project, the Center for Climate and Energy Solutions, and the Center for Climate and Security. The 2007 report by CNA's Military Advisory Board (MAB; a panel of retired generals and admirals), which articulated the security implications of climate change, is considered by many to be a landmark event in the U.S. climate change conversation (Catarious et al. 2007); the MAB released an updated report in 2014, refining their findings from the 2007 report, stressing the security implications of climate change, and urging action on the part of the U.S. government (Goodman 2014). Within the United States, the federal government has not established a uniform, consistent policy regarding climate change and security, and both the academic and policy communities continue to struggle to establish an effective research agenda (and policy based on effective research) related to climate change and security. The Syrian civil war presents an opportunity to examine the emergent security landscape as it relates to environmental security, including possible connections between climate change and security.

3 Case Study: The Syrian Civil War

Many factors contributed to the outbreak of armed conflict in Syria, such as widespread discontent with the Assad regime, sectarian friction, high unemployment, the success of Arab Spring movements in changing regimes in nearby states, and a response to a heavy–handed government crackdown on the early Arab Spring protesters in Syria. Another factor that may have contributed to the conflict, but that is not often discussed, is the long term (2006–2010) drought in agricultural regions of northeast Syria. Some scholars have argued that the drought is attributable to climate change, and suggest, therefore, that anthropogenic climate change was a factor in triggering the Syrian civil war. Such claims have been echoed by some policy makers, media, and commentators. Others critique such claims, arguing that the effects of the drought, especially drought–induced migration, have been exaggerated, that it is not possible to attribute the drought to climate change (natural or anthropogenic), or that it is simply inaccurate to attribute the civil war to the drought in light of other, more significant socio-political factors. This section begins with a

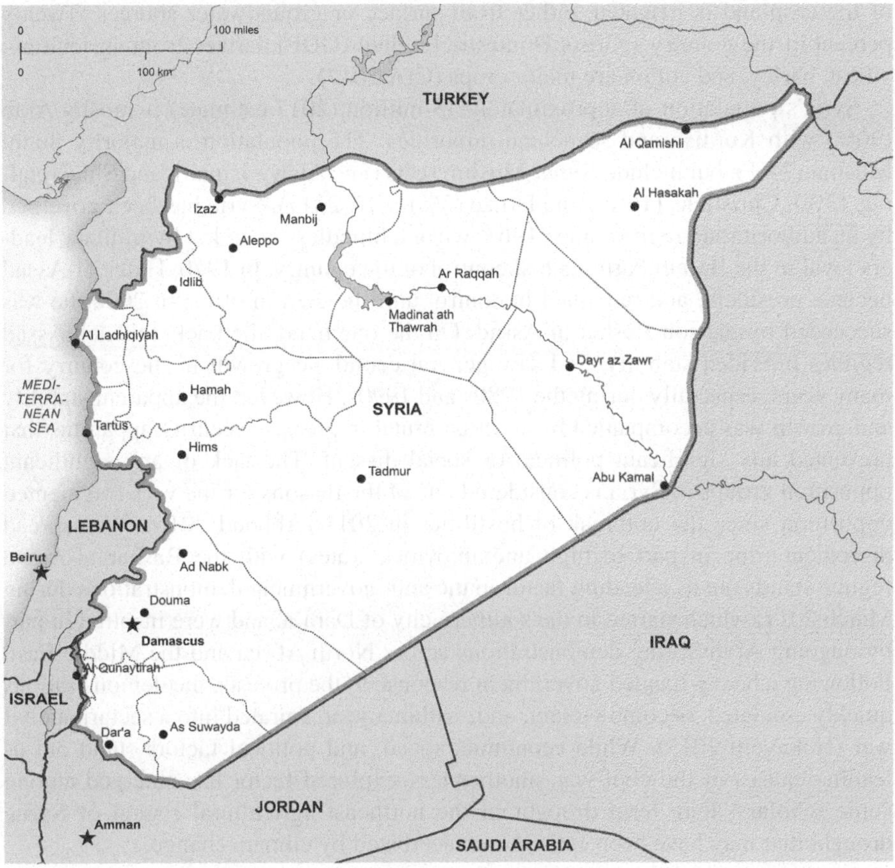

Fig. 1 Syria, 2017 (www.onestopmap.com)

brief background of Syria and its civil war. Then the claims linking climate change to the civil war are explored.

3.1 *Background*

Syria is a Middle Eastern country of approximately 185,000 square kilometers (about 1.5 times the size of Pennsylvania) in the eastern Mediterranean, or Levant region (Fig. 1). The country consists of semiarid and desert plateau, with a narrow coastal plain and mountains along the Mediterranean Sea in the west. Summers are hot and dry, with weather patterns dominated by subtropical high pressure. Winters are mild, with most of the country's annual precipitation falling from December to February, brought by midlatitude cyclones. About 75% of Syria's land is agricultural (30% arable or permanent cropland, 45% pasture). Approximately one quarter of the cropland is irrigated, either from surface or groundwater sources. Twenty percent of the country's Gross Domestic Product (GDP) derives from agriculture; wheat, barley, and cotton are major crops (CIA 2017).

Syria's population of approximately 18 million (2017 estimate) is mostly Arab (90%) with Kurdish and Armenian minorities. The population is majority Sunni Muslim (74%), but includes other Muslim sects (i.e., Alawi, Ismaili, and Shia, totaling 13%), Christians (10%), and Druze (3%) (CIA 2017). Syria has been governed by an authoritarian regime since 1963, when a bloodless coup led by military leaders loyal to the Ba'ath Party gained control of the country. In 1970, Hafez al-Assad became president, and remained in control until he died in office in 2000; he was succeeded by his son Bashar al-Assad. On the one hand, the back–to–back Assad regimes provided stability, and saw general economic growth in the country for many years, especially during the 1980s and 1990s. However, the apparent stability and growth was accompanied by an often brutal, repressive security apparatus that prevented any significant political or social dissent. The lack of any significant opposition groups in Syria is considered one of the reasons for the very fragmented opposition since the outbreak of hostilities in 2011 (Abboud 2016). Widespread discontent (due in part to high unemployment rates) with the Bashar al-Assad regime stands out as a leading factor in the anti–government demonstrations during March 2011, which started in the southern city of Dara'a, and were inspired in part by ongoing Arab Spring demonstrations across North Africa and the Middle East. Following a heavy-handed government response to the protests, the demonstrations quickly escalated, become violent, and, within a year, spiraled into a sectarian civil war (Hokayem 2013). While economic, social, and political factors stand out as leading causes of the civil war, another less–explored factor has emerged among some scholars: long term drought in the northeast agricultural region of Syria, drought that may have been caused or exacerbated by climate change.

3.2 The Role of Climate Change

A debate about the role that climate change, drought, and migration may have played in the Syrian civil war has developed among a small group of scholars. Those arguing that anthropogenic climate change played a role in the outbreak of the civil war make three claims. First, the drought in northeast Syria from 2006 to 2010 was caused by, or made worse by, human-induced climate change. Second, large numbers of Syrians affected by the drought migrated to already crowded cities in the country. And finally, the drought–induced migration resulted in overcrowding in major population centers, worsening already poor living conditions, employment opportunities, and access to basic services, in turn fueling anger and discontent among both residents and migrants. In this section, each of these points (and respective counterpoints) is presented.

There is no dispute that the drought in northeast Syria from 2006 to 2010 was among the most severe on record—both sides of the debate agree on this point

(Selby et al. 2017). But can the drought, or its severity, be attributed to anthropogenic climate change? Proponents of the climate change link, including Gleick (2014), Werrell et al. (2015), and Kelley et al. (2015) all point to a single source, a study by Hoerling et al. (2012) that states the warming and drying trends across the Mediterranean region over the past century are explained by anthropogenic climate change, not natural processes alone. Selby et al. (2017) reanalyzed the data, and came to different conclusions. First, while there are long term warming and drying trends across the Mediterranean region, the trends at more localized spatial levels are varied, with some areas becoming drier and other areas becoming more wet. Second, the uncertainty associated with climate models used by Hoerling et al. (2012) and Kelley et al. (2015) is too great to conclusively attribute the drought in northeast Syria to anthropogenic climate change. To do so is "*to confuse probabilistic and deterministic claims*" (Selby et al. 2017, 235). The weight of the evidence seems to suggest that attributing the severe drought in Syria in the late 2000s to anthropogenic climate change is tenuous.

The second claim linking climate change to the civil war is that millions of Syrians affected by the drought in northeast Syria opted to migrate, mostly from rural, drought affected areas to already overcrowded cities in Syria. Proponents of this claim state that between 1.5 and 2 million Syrians migrated in the late 2000s as a result of lost livelihood because of the long term drought (Gleick 2014; Kelley et al. 2015; Werrell et al. 2015); this number seems to have been cited from a single Syrian government official in 2009 (IRIN 2009). Selby et al. (2017) acknowledge that the severe drought did cause loss of livelihood, and some chose to migrate, but they cast doubt on the numbers who opted to migrate, as well as their motives for migrating. Official U.N. and Syrian government reports of migrants from the northeast region in the late 2000s estimate 40,000–60,000 families in the 2008–2009 timeframe. Even if that number of families migrated for multiple years, the number falls far short of the frequently cited 1.5–2 million migrants. Additionally, other factors may have motivated those who did migrate, including economic liberalization policies (Selby et al. 2017). Although millions of Syrians have been displaced, both internally and externally, as a result of the civil war, there remains doubt about causal role of drought-induced mass migration in northeast Syria before the war.

The final claim linking climate change to the civil war is that the aforementioned mass migration and subsequent overcrowding pushed the affected populations to, or close to, a breaking point. Proponents suggest that drought-affected rural to urban migrants put additional strain on Syria's already crowded cities, contributing to public demonstrations of grievances (Werrell et al. 2015; Gleick 2014), and this rapid, mass migration worsened a number of other factors in the unrest (including unemployment, corruption, and inequalities)(Kelley et al. 2015). Opponents of this claim cite a lack of evidence supporting the assertions about population pressures caused by migrants to Syrian cities, and suggest that those who did migrate from northeast Syria in the late 2000s were not involved in the initial protests and unrest in 2011 (Selby et al. 2017). Further research on this claim, including more extensive social science research, might shed additional light on the role that migrants played in the initial uprisings and ongoing conflict.

To summarize, debate has emerged among a small group of scholars over the role of climate change, drought, and migration as causal or contributing factors in the outbreak of the Syrian civil war. Proponents of the linkages claim that climate change caused or exacerbated drought in northeast Syria from 2006–2010, that the severe drought led to the migration of 1.5–2 million rural Syrians to cities, and that the pressure of these migrants contributed to the unrest in 2011, which soon led to the outbreak of hostilities across the country. Opponents dispute each of these claims, suggesting that while there was a severe drought in the northeast of Syria, it cannot be conclusively attributed to climate change, nor did it cause migration on the scale suggested by proponents, and those who did opt to migrate did not play a significant role in the unrest and ensuing hostilities in 2011–2012.

4 Conclusions

The role that climate change or drought played in the outbreak or exacerbation of the Syrian civil war may be difficult to prove with any degree of certainty. Nonetheless, there are lessons that we can learn from the Syrian civil war that relate to environmental security. First, water and climatic conditions play an important role in the economic and social conditions of the Middle East broadly, and Syria specifically. In many instances, at both the regional and local levels, past water use practices are unsustainable. Gleick (2014) suggests three options to move toward more sustainable water use: improve water efficiency in agriculture, improve monitoring and management of groundwater resources, and continue to work toward more comprehensive international agreements on use of the region's rivers that flow across state borders. Second, violent conflict has a significant effect on the environment. The impact of violent conflict on the environment has long been a subset of environmental security, and the Syrian civil war provides another, albeit tragic, example. In Syria, one of the most significant environmental impacts of the war is on agriculture. From 2011–2016, the cost of damage and loss in the country's agricultural sector was an estimated $16 billion, including $3.2 billion in damage to irrigation and other agriculture infrastructure, with 60% of households reporting significant damage (FAO 2017). As conditions permit, Syria, with significant assistance from the international community, will begin the rebuilding process. Though expensive, with rebuilding comes opportunities to introduce more sustainable agricultural and other environmental infrastructure and practices. Third, precise language and rigorous research matter, especially when communicating environmental security-related research to policy makers, the media, and the general public. The previously discussed debate surrounding the role of climate change in the Syrian civil war represents a healthy dialog among scholars that will contribute to the rigor of future research in this area. Some members of the media and the policy community have picked up on elements of this research, without seeking to understand context or limitations of the findings. Selby et al. (2017) rightly caution that, given the "*urgency of the climate change challenge and the contestation around it, plus the mass media's preference for striking, overblown stories,"* researchers should avoid exaggerating climate-conflict links, and that scholars should draw on cross-disciplinary expertise for such complex problems (241).

Looking forward, the Syrian civil war provides opportunity for future environmental security research. Areas that are ripe for research include a more comprehensive assessment of the emergent effects of human insecurity (including drought, food insecurity, and migration) on local, regional and international security interests, as well as more effective methods for bridging the science-policy-general public gaps on complex issues such as climate change and conflict. The environmental impacts of the Syrian civil war, especially the effects on agriculture, land use, and water resources represent another area that calls for more research. Finally, further study of environmental security implications of the Syrian civil war demands employment of rigorous cross-disciplinary research to study, understand, and communicate complex socio-environmental challenges and opportunities.

Literature Cited

Abboud SN (2016) Syria. Polity Press, Malden

Adger WN, Pulhin J (2014) Human Security. In: Field CB, Barros VR, Dokken DR, Mach KJ, Mastrandrea MD, Bilir TE, Chatterjee M, Ebi KL, Estrada YO, Genova RC, Girma B, Kissel ES, Levy AN, MacCracken S, Mastrandrea PR, White LL (eds) Climate change 2014: impacts, adaptation and vulnerability. Cambridge University Press, New York, pp 755–792

Barnett J (2003) Security and climate change. Glob Environ Chang 13(1):7–17

Catarious DM, Filadelfo R, Gaffney H, Maybee S, Morehouse T (2007) National Security and the threat of climate change. CNA, Alexandria

Central Intelligence Agency (2017) The World Factbook: Syria. https://www.cia.gov/library/publications/the-world-factbook/geos/sy.html. Accessed 28 Jan 2018

Food and Agriculture Organization of the United Nations (2017) Counting the cost: agriculture in Syria after six years of crisis. Food and Agriculture Organization of the United Nation, Roma

Fund for Peace. Fragile States Index (2017) http://fundforpeace.org/fsi/. Accessed 15 Jan 2018

Gleick P (2014) Water, drought, climate change, and conflict in Syria. Weather, Climate, Soc 6:331–340

Goodman S (2014) National Security and the accelerating risks of climate change. CNA Military Advisory Board, Alexandria, VA

Hoerling M, Eischeid J, Perlwitz J, Quan X, Zhang T, Pegion P (2012) On the increased frequency of Mediterranean drought. J Clim 25:2146–2161

Hokayem E (2013) Syria's uprising and the fracturing of the Levant. Routledge, New York

Hsiang SM, Burke M, Miguel E (2013) Quantifying the influence of climate on human conflict. Science 341(6151):1235367

IRIN (2009) Drought response faces funding shortfall. http://www.irinnews.org/news/2009/11/24/drought-response-faces-funding-shortfall. Accessed 1 Mar 2018

Kelley CP, Mohtadi S, Cane MA, Seager R, Kushnir Y (2015) Climate change in the fertile crescent and implications of the recent Syrian drought. Proc Natl Acad Sci 112(11):3241–3246

Nordas R, Gleditsch NP (2007) Climate change and conflict. Polit Geogr 26(6):627–638

Parry M, Canziani OF, Palutikof JP, van der Linden PJ, Hanson CE (eds) (2007) Climate change 2007: impacts, adaptation, and vulnerability: contribution of Working Group II to the fourth assessment report of the Intergovernmental Panel on Climate Change. Cambridge University Press, New York

Raleigh C, Urdal H (2007) Climate change, environmental degradation and armed conflict. Polit Geogr 26(6):674–694
Selby J, Dahi O, Frohlich C, Hulme M (2017) Climate change and the Syrian civil war revisited. Polit Geogr 60:232–244
Sorenson D (2016) Syria in Ruins. Praeger Security International, Santa Barbara
US Congress (2007) National Defense Authorization Act for fiscal year 2008
US Congress (2017) National Defense Authorization Act for fiscal year 2018
Werrell C, Femia F, Sternberg T (2015) Did we see it coming?: state fragility, climate vulnerability, and the uprisings in Syria and Egypt. SAIS Rev Int Aff 35(1):29–46

Index

F. Galgano (ed.), *The Environment-Conflict Nexus*, Advances in Military Geosciences, https://doi.org/10.1007/978-3-319-90975-2

The manufacturer's authorised representative in the EU is Springer Nature Customer Service Centre GmbH, Europaplatz 3, 69115 Heidelberg, Germany. If you have any concerns regarding our products, please contact ProductSafety@springernature.com

Printed and bound by CPI Group (UK) Ltd, Croydon, CR0 4YY
30/09/2026
02225658-0001